# Modelling Hydrology, Hydraulics and Contaminant Transport Systems in Python

# Modelling Hydrology, Hydraulics and Contaminant Transport Systems in Python

Soumendra Nath Kuiry

Dhrubajyoti Sen

CRC Press
Taylor & Francis Group
Boca Raton London New York

CRC Press is an imprint of the
Taylor & Francis Group, an **informa** business

First edition published 2022
by CRC Press
6000 Broken Sound Parkway NW, Suite 300, Boca Raton, FL 33487-2742

and by CRC Press
2 Park Square, Milton Park, Abingdon, Oxon, OX14 4RN

© 2022 Taylor & Francis Group, LLC

First edition published by CRC Press 2022

CRC Press is an imprint of Taylor & Francis Group, LLC

ISBN: 978-0-367-25578-7 (hbk)
ISBN: 978-1-032-12989-1 (pbk)
ISBN: 978-0-429-28857-9 (ebk)

DOI: 10.1201/9780429288579

Typeset in Times
by SPi Technologies India Pvt Ltd (Straive)

# Contents

# Preface

The need for a book on the use of numerical techniques in environmental and civil engineering was felt during the course of our teaching allied subjects at the senior undergraduate and graduate levels. Although the first- and second-year level undergraduate syllabi on engineering mathematics deal extensively with a variety of modelling tools and solution techniques for handling different types of scientific and engineering equations, the specific use of numerical techniques in solving problems of hydraulics, hydrology, and contaminant transport is found lacking in the graduation level courses in most colleges.

Nonetheless, there has been an ever-growing interest among the student community on the use of the open-source programming language Python, especially for data handling and data-based computations, thanks to the availability of a wealth of data from the environmental, especially, hydro-meteorological fields. In this book, we intend to introduce the reader to the rudiments of the application of numerical techniques to a variety of problems encountered in the field of water-environment through the use of coding. Preference in using the programming language Python has mostly been for the reason that it has become, at least at the time of writing this book, quite popular among students in addition to it being open source with a wealth of helpful online resources. It is possible that over the coming years, yet another programming language may become popular just as has been witnessed over the past few decades with the popularity of scientific coding shifting from FORTRAN to C/C++, and then to computing platforms like the MATLAB®. Nevertheless, it is hoped that the section of the book demonstrating the use of numerical techniques in finding solutions to the physical problems of hydrology, hydraulics, and transport would still remain helpful to the future reader.

On the whole, it must be emphasized that this book is meant to be used more as a teaching aid for the senior undergraduate and graduate classes and for them to gain an understanding of the physical processes in the world of the water-environment through computer-based hands-on problem solving. It is hoped that on gaining an interest in the field of modelling, the keen reader would be inspired to venture for more accurate solution techniques or start using better and efficient coding platforms for advanced research projects.

The book is divided into seven chapters, of which Chapter 1 gives an overview of the models and their types in the field of hydrology, hydraulics, and contaminant transport, which have been the focus of this book. Chapter 2 elaborates on the solution of non-linear simultaneous equations, taking up examples from the field of simple hydraulic and water-engineering problems. Chapter 3 provides examples of ordinary differential equations from the field of the water-environment. The next three chapters, Chapters 4, 5, and 6, discuss the application of numerical techniques to surface flows, subsurface flows, and contaminant transport, respectively. The final chapter (Chapter 7) serves as an exposure to the rudimentary quantitative analysis of different data sets, as encountered in the field of the hydro-environment. This chapter, therefore, differs from the others in not dealing with physics-based models but, at the same time, exposes the reader to a variety of quantitative methods commonly used in analysing observational data. In fact, some of the techniques demonstrated in Chapter 7, though simple, may be used to process a wide body of field and laboratory data and then apply these for calibrating and validating one or more of the physical models introduced in the previous chapters.

Finally, the authors would like to thank the CRC Press for providing them the opportunity and encouragement to script their ideas that otherwise would probably never have seen the light of the day. Dr. Gagandeep Singh and Mr. Lakshay Gaba of the CRC Press are to be specially acknowledged for constantly monitoring the editorial process and bringing the project to fruition in a short time.

Soumendra Nath Kuiry

Dhrubajyoti Sen

MATLAB® is a registered trademark of The Math Works, Inc.
For product information, please contact:
The Math Works, Inc.
3 Apple Hill Drive
Natick, MA 01760-2098
Tel: 508-647-7000
Fax: 508-647-7001
E-mail: info@mathworks.com
Web: http://www.mathworks.com

# About the Authors

**Soumendra Nath Kuiry** is a faculty at the Indian Institute of Technology Madras, with expertise in developing computational techniques in the different processes of free surface flows. Specializing in flood modelling, his research interests extend over simulations of tsunami wave propagation and dam break phenomena, modelling of storm surges due to cyclones and simulation of sediment transport in rivers, estuaries, and coasts.

**Dhrubajyoti Sen** is a faculty at the Indian Institute of Technology Kharagpur, with research interests in experimental and numerical modelling of surface flows, dam break incidents, storm surges, and contaminant transport. His interests in practical projects have also led him to explore the application of digital methods and electronic sensing technologies in environmental sciences and hydraulic engineering.

# 1 Introduction to Modelling in Hydrology, Hydraulics, and Contaminant Transport

## 1.1 EXAMPLES OF DIFFERENT TYPES OF MODELS IN WATER SYSTEMS: DETERMINISTIC, STOCHASTIC, DATA-BASED, AND OTHERS

Water systems vary widely in applications and the present text is aligned mostly towards the hydraulic and hydrologic phenomena as dealt with in civil, environmental, or agricultural science and engineering disciplines. The phenomena discussed include those from the surface and subsurface phases of the hydrologic cycle, and open channel flows through natural or engineered structures. From its initial touchdown till its eventual dispersal to the ocean, the rainwater collects and transmits pollutants, as it flows through the different pathways. The text therefore also discusses the simulation techniques for predicting the fate and transport of the water pollutants in channels, shallow water bodies, soil-seepage, and groundwater flows.

The hydrologic surface and subsurface flows are also related to conventional hydraulics and may be approximated through suitable mathematical relations or equations. The different chapters of this book demonstrate the numerical techniques by which these equations may be solved on a personal computer. In the present age of "big-data", there is an increasing trend of modelling the hydrologic and hydraulic variables through data analysis – such as by using stochastic or data-driven techniques. These tools are available through various platforms and help in obtaining the "bigger picture" of the environmental processes occurring around us. However, on certain smaller scales, it is still necessary to understand the physics behind a hydraulic phenomenon and predict its behaviour using compatible models. For example, atmospheric models and data analytics may help predict future rainfall over a catchment but precise delineation of the flood-inundated areas may only be possible by solving the free-surface flow equations using suitable computational methods. Similarly, computing the extent of surge that may move up a tidal river under predicted future conditions of sea level rise may only be possible using numerical models to solve the flow equations with the help of a computer. For predicting the movement and extent of spread of a contaminant along with the flow of water,

DOI: 10.1201/9780429288579-1

whether over the earth's surface or below, would likewise require solving the appropriate equations numerically on a computer.

## 1.2  CHOOSING A NUMERICAL APPROACH FOR FLOW AND TRANSPORT MODELLING

Many of the physical processes of environmental and engineered flows may broadly
be described mathematically by one simple equation, a set of simultaneous equations, or differential equations. In each of these types of equations, the different flow
variables are interrelated in terms of the rate of change of one or several variables as
a function of time and/or space variables. Under certain simplified conditions, the
equation(s) may be solved analytically, and may not require a numerical solution to
solve using computers. Some of such examples include the formulae for evapotrans-
piration and infiltration, expressed in terms of the independent variables. However, in
many practical situations, the unique geometry of the flow domain may make it dif-
ficult for applying analytical techniques, as the processes can only be described by
non-linear partial differential equations. As a result, only numerical solutions of the
equations offer the feasible means for obtaining the desired results. Oftentimes, the
variation of flow with time may also demand the application of specific numerical
techniques. It is important to note that numerical solutions, though producing approximate results, may yield better accuracy either by refining the equations themselves
or by adopting more accurate numerical techniques. Quite often, however, simple
approximations may also yield reasonably acceptable results, which may help in
understanding the occurrence of a particular phenomenon or aid in taking a rapid
decision. The contents of this book do not venture into very accurate numerical
approximations of the different hydraulic phenomena as encountered in the natural
or built environment. Rather, they attempt to demonstrate the general methods that
may be selected for tackling a given problem and obtaining an initial solution. Further
improvement and refinement of the models are possible by expanding on the building
blocks demonstrated here on the application of numerical techniques to different
geophysical and engineered flows.

## 1.3  PYTHON AS THE PREFERRED PROGRAMMING PLATFORM

For solving the numerically approximated equations, one needs to write computer
programs or codes, which may be run on a suitable computational platform. The
traditionally used computer language for scientific code writing since the middle of
the last century had been FORTRAN, and sometimes BASIC and others. However,
these were taken over by the end of the century by C/C++, and to some extent by
Java. However, the last century also saw the rise of computational platforms like
MATLAB, or other similar interpretable programming platforms, which became
popular among researchers and still remain so. For this book, however, we have chosen the language Python, which appears to have been gaining popularity lately, especially among the student and academic community. Since the purpose of this text is
more pedagogical than a research exposition, we have also chosen Python as the
preferred coding language because of its relatively easy learning curve. Python is

also a completely open-source computing platform and is rather easy to install and start working. An added advantage of using Python as the coding language is that it provides readily accessible functions for plotting graphs. Thus, it may be used to display and save graphical outputs of the variables which are used as inputs or those which are obtained from running a code. Codes written in FORTRAN or C/C++ first require the variables to be stored in files and then plotted graphically using another software package, such as the Paraview. Although Python may have become popular for its varied applications in non-scientific and scientific but non-numerical applications, Python has also been demonstrated for use in numeric computations, as proved by the appearance of books on the subject over the past decade. Although one may use Python to write codes on the go, such as on interpretable computation platforms, in this book we shall encourage the readers to write scripts (equivalent to computer codes) and then run the codes using the Python command. This is definitely required for longer programs since the errors in the script may be corrected or modified for future reuse. Further, since there are several books and online resources guiding the installation of the Python computing platform, it is not discussed in this book.

## 1.4 PEDAGOGICAL EMPHASIS

This book is primarily intended for the students of hydraulics, hydrology, environmental, and water resources engineering who may like to start writing their own codes for the problems at hand. Though nowadays many open-source software packages are available for the accomplishment of different tasks in the fields described, quite often, especially when the problem is simple in terms of geometry and other physical properties, it is enough to write a simple computer program and obtain the solution. The same program may be later improved and adapted to more complicated inputs and geometries. Since the book also discusses the popular numerical schemes used for solving problems in the field of hydrology, hydraulics, and transport of contaminants by flowing water it may be used for teaching a semester-long course on numerical methods for the students of such specializations. The codes are not written in the most optimum way and may be modified by the student in order to make it more efficient in terms of memory management or time of execution. Similarly, the student may experiment with other advanced forms of graphical outputs, including animations.

## 1.5 TYPES OF MODELS TREATED IN THE BOOK

The models presented in this book on hydrology, hydraulics, and contaminant transport are demonstrated for the following physical processes:

1. Surface runoff generation by rainfall
2. Flows in one-dimensional open channels
3. Depth change in reservoirs from inflows and outflows
4. Flows in shallow lakes and water bodies
5. Flows in pipe networks
6. Shallow groundwater flow, considered two-dimensional in the horizontal plane

7. Saturated seepage flow through soils in the vertical two-dimensional plane
8. Contaminant transport in one-dimensional open channel flows, shallow two-dimensional surface flows, and seepage flows in soils.

The equations involved are of the following types:

1. Non-linear equation
2. Simultaneous linear and non-linear equations
3. Ordinary differential equations
4. Partial differential equations

The numerical techniques discussed are:

1. The Newton–Raphson method of finding the roots of non-linear equations
2. Solution of simultaneous linear equations using the Gaussian elimination method
3. Solution of simultaneous non-linear equations by a combination of the above
4. Solution of ordinary and partial differential equations using the method of finite differences

This book is not intended to elaborate on the numerical techniques, as many useful texts are widely available, some of these being listed in the bibliography. Nor is this book meant to be a programming guide for the language Python, for which again several books and web resources are available. The reader is thus encouraged to proceed only after acquiring a basic knowledge of coding in Python in order to find the book useful. Nonetheless, it is emphasized that this book should be found useful to the science and engineering students who have started exploring the exciting world of coding and modelling physical processes, especially those involving the flow of water and movement of contaminants in the natural and built environments.

# 2 Non-Linear and Simultaneous Equations

This chapter introduces a variety of physical processes – sometimes called as systems – in hydraulics, and water resources engineering and are described equivalently in the mathematical form as a single or a set of equations. If an equation is possible to be rearranged in terms of the unknown variable explicitly, we may obtain an answer without resorting to advanced computational methods. However, if the equation is non-linear in terms of the unknown variable, then it may not be possible to write an explicit expression for evaluating the variable. In such cases, root-finding algorithms – like the Newton–Raphson Method – may have to be used. An extension of the single equation involving one unknown variable is the system of say, $n$, equations in terms of $n$ independent variables. Here too, the $n$ equations may be solved for the $n$ unknowns by standard methods of linear algebra if the system of equation is linear, that is if the variables are not arranged in powers greater than one. Otherwise, techniques such as the Newton–Raphson may be used by extending it for $n$ unknowns. This chapter demonstrates the solutions for some problems of these kinds and provides computer programs in Python for implementing the solution algorithms.

## 2.1 EXAMPLES OF NON-LINEAR FUNCTIONS

Assuming that finding solutions to linear equations in one-variable does not require the knowledge of any special technique, we start with a few examples of non-linear equations (in one-variable) for problems encountered in hydraulic engineering.

### 2.1.1 NORMAL DEPTH OF FLOW IN A TRAPEZOIDAL CHANNEL

An example of a non-linear equation encountered in the field of hydraulics is the finding of normal depth ($y_n$) of the flow taking place through a long trapezoidal channel (Figure 2.1). Such a flow depth occurs at the uniform flow region, much upstream from the influence of any cause for non-uniformity, such as a weir.

The normal depth $y_n$ may be obtained by solving the following equation (Chaudhry, 2008):

$$Q = \frac{1}{n} AR^{2/3} S_0^{1/2} \tag{2.1}$$

where $Q$ is the steady-state discharge passing through the channel, $n$ is the Manning's roughness coefficient, $A$ and $R$ are the cross-sectional area and wetted perimeter, respectively, both of which are functions of the depth of water, and $S_0$ is the

DOI: 10.1201/9780429288579-2

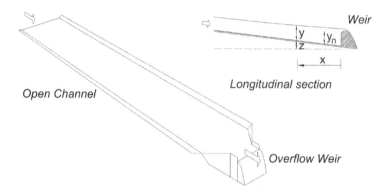

**FIGURE 2.1**  Long trapezoidal channel, showing normal depth occurring far upstream from the influence of weir.

longitudinal slope of the channel. Accordingly, since $A$ and $R$ are both functions of the normal depth $(y_n)$, and together make the equation non-linear, Equation (2.1), is regarded as an implicit equation in terms of the unknown (that is, $y_n$), and needs to be iteratively solved using a suitable numerical method since direct analytical solution is not readily available.

## 2.1.2   HEIGHT AND VELOCITY OF A SURGE WAVE

Another situation of a non-linear equation, also taken from the field of hydraulic engineering, is the computation of the height of a surge wave occurring in a one-dimensional channel due to a sudden increase or decrease of the flow at one of its ends. Figure 2.2 shows two such cases, one produced by a decrease of discharge at the downstream end (Figure 2.2a), and the other produced by an increase in discharge at the upstream end (Figure 2.2b). Both figures depict what is known as positive surge waves, whereby the depth of the water behind the travelling wave is greater than that in front of it. The first case may occur in a hydropower canal in which a downstream gate, controlling the water entering the turbines, is operated suddenly to reduce the discharge. The other may be possible in the exit channel of a hydropower generating unit where the discharge is suddenly increased due to the commencement of a turbine runner. Another category of surge waves may occur in a channel with a gate holding back a reserve of water, which is suddenly withdrawn. A negative surge is then created in which the flow depth behind the wave travelling up the reservoir is smaller than that in front of it.

For the positive surge shown in Figure 2.2(b), the velocity of the moving surge $(v_w)$ in a rectangular channel is given in terms of the upstream and downstream depths $(y_2$ and $y_1$, respectively) and corresponding velocities $(v_2$ and $v_1$, respectively) as follows (Chaudhry, 2008):

$$v_w = \frac{v_2 y_2 - v_1 y_1}{y_2 - y_1} \qquad (2.2)$$

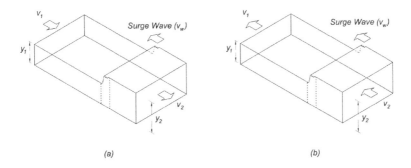

**FIGURE 2.2** Surge waves generated by: (a) reduction of outflow discharge or velocity ($v_2$), and (b) increase in inflow discharge or velocity ($v_2$). Both are positive surges.

Another relation, which connects the upstream and downstream depths but does not include the surge velocity, is given below and is useful in finding the unknowns from the given conditions in a problem:

$$\left(v_1 - v_2\right)^2 = \frac{g\left(y_1 - y_2\right)}{2y_1 y_2}\left(y_1^2 - y_2^2\right) \tag{2.3}$$

While solving Equation (2.3), the unknowns are generally the variables $y_2$ and $v_2$ since the other parameters are usually known from the initial conditions. However, since these two variables are related to one another through the continuity equation, the equation becomes implicit with either $y_2$ or $v_2$ as the unknown. Hence, a suitable numerical technique needs to be employed to find the unknown variable, as discussed in subsequent sections.

### 2.1.3 DEPTH OF FLOW IN A CONSTRICTED AND RAISED CHANNEL SECTION

The waterway for flows taking place through bridge culverts is often reduced in comparison to the width of the unrestricted channel. This increases the discharge concentration through the culvert, leading to an increased velocity and corresponding changes in the water surface elevation. Quite often, the floor of the culvert is also raised, which results in varying the water level further. For a given upstream flow discharge and known amounts of width constriction and bed rise, the problem of determining the water surface elevation at the culvert requires solving a non-linear equation as shown below.

Consider a horizontal rectangular open channel, locally constricted in width and with a raised bed, passing a given discharge, $Q$ (Figure 2.3). The unrestricted width of the section is $b_1$ while that of the constriction is $b_2$. The height of the hump at the contracted section is $\Delta z$ above the level of the bed of the wider section. The depth of flow in the upstream section is $y_1$ and it is required to find the depth $y_2$ over the hump.

Since the changes in depths take place over a short length of the channel and if the contraction is largely smooth, we may ignore any energy loss and assume that the

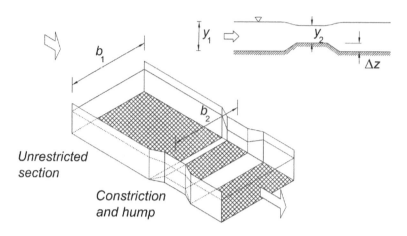

**FIGURE 2.3**    Steady flow passing through an open channel with constriction and hump.

specific energies between a point on the upstream and another over the hump are nearly equal. The specific discharges, that is the discharges per unit width, $q_1$ and $q_2$, respectively, on the upstream and over the hump may be first evaluated using the continuity equation, since the total discharge $Q$ and the width of the two sections are known. The total energies at the wider and narrower sections, considering the original bed level as datum and assuming no head loss, may then be equated to obtain the following equation:

$$y_1 + \frac{q_1^2}{2gy_1^2} = y_2 + \frac{q_2^2}{2gy_2^2} + \Delta z \tag{2.4}$$

The terms on the left of Equation (2.4) being known, simplifies to a third-degree equation in terms of the unknown depth $y_2$, which needs to be solved.

## 2.2   SYSTEM OF EQUATIONS

A set of $n$ equations, containing $n$ unknown variables, commonly occurs while solving simultaneous equations. Some representative examples of such "system of equations", taken from the fields of hydraulic engineering and hydrology, and requiring the solution of the equations simultaneously, are presented below.

### 2.2.1   SYSTEM OF REACTORS – STEADY-STATE ANALYSIS

Well-mixed lakes receiving polluted water from a source with simultaneous outflow from another end may be likened to continuous-flow stirred tank reactors (Chapra, 1997). Figure 2.4 shows a hypothetical system of five such interconnected lakes represented by equivalent tank reactors, with various inflow and outflow routes. The example has been adapted from a similar example presented by Chapra and Canale

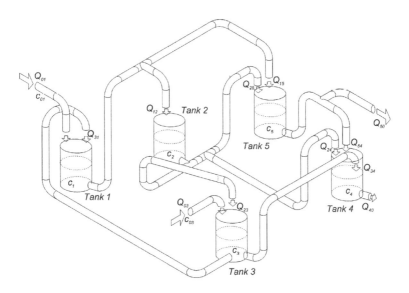

**FIGURE 2.4** A combination of five well-mixed interconnected lakes represented by equivalent tank reactors and flow conduits (adapted from Chapra and Canale, 2021).

(2021) and assumes a steady-state flow-rate situation. The tanks are numbered serially from 1 to 5 and a configuration of pipes are assumed that supply flows contaminated with a conservative pollutant to some of the reactors. The flow rate is denoted by $Q_{0x}$ while the concentration of the contaminant brought along with the flow is denoted as $c_{0x}$. The subscript $x$ represents the tank number and the 0 preceding $x$ indicates that the flow is entering the system from outside. The pipe arrangements are assumed to permit some tanks to let the flow out of the system and the corresponding out-flowing discharges are denoted as $Q_{x0}$. Some of the tanks receive flow (along with some amount of the dissolved substance) from other tanks and although the flow rates in the connecting pipes are assumed as known, concentration of the liquids in the pipes or in the tanks (denoted by $c_n$, where $n$ is the tank number) are as yet unknown. This steady-state problem can be solved by setting up a mass balance equation for the solute in each tank, or the lakes that they represent. The discharges between the tanks are denoted as $Q_{xy}$, where $x$ and $y$ in the subscript denote the donor and receiving tanks, respectively.

Thus, for the configuration shown in Figure 2.4, the mass balance relations of the contaminant in each tank under the assumed steady-state conditions may be expressed by the following set of equations:

$$\text{Tank1}: Q_{01}.C_{01} + Q_{31}.C_3 = Q_{12}.C_1 + Q_{15}.C_1$$
$$\text{Tank2}: Q_{12}.C_1 = Q_{25}.C_2 + Q_{24}.C_2 + Q_{23}.C_3$$
$$\text{Tank3}: Q_{03}.C_{03} + Q_{23}.C_2 = Q_{31}.C_3 + Q_{34}.C_3 \qquad (2.5)$$
$$\text{Tank4}: Q_{24}.C_2 + Q_{34}.C_3 + Q_{54}.C_5 = Q_{40}.C_4$$
$$\text{Tank5}: Q_{25}.C_2 + Q_{15}.C_1 = Q_{54}.C_5 + Q_{50}.C_5$$

The five linear equations represented by Equation (2.5) need to be solved simultaneously for the five unknown concentrations in the five tanks ($c_1$ to $c_5$) using standard procedures like the Gaussian Elimination, as explained subsequently.

## 2.2.2 STEADY-STATE DISTRIBUTION OF FLOW IN PIPE NETWORKS

In water supply engineering, it is often required to compute the flow through a network of pipes, the source being one or more elevated water tanks (or pumps) and the demands at the different junctions are assumed to be known. An example is shown in Figure 2.5, which shows three constant-head tanks, $a$, $b$, and $c$ of known pressure heads $Ha$, $Hb$, and $Hc$, respectively. There are three connecting pipes with known length, diameter, and roughness properties, which are assumed to meet at a junction. The pressure head at this junction is denoted by $Hd$, which is as yet not known.

Since the directions of flow in the pipes are not known initially, they are assumed as shown in Figure 2.5. It is then possible to write an equal number of equations as there are unknowns by making use of the continuity equation at the junction and equating the total energy heads between the tanks. Since there are no closed loops in the example, the energy equivalence equations are expressed for the "open" loops connecting pairs of tanks at a time. However, the number of independent equations possible this way is always one less than the number of tanks. This is because there is one equation that may be derived by manipulating the others, and is thus redundant. However, we may derive the last equation by making use of the continuity principle at the junction of the pipes. Denoting the discharges in the three pipes as $Q_1$, $Q_2$, and $Q_3$, respectively, the continuity equation takes the following form:

$$Q_1 = Q_2 + Q_3 \qquad (2.6)$$

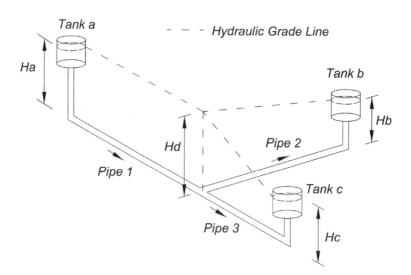

**FIGURE 2.5** Flow distribution in a network of three pipes connecting three overhead water reservoirs.

Next, by balancing the energy heads between tanks $a$ and $b$, we obtain the following expression:

$$H_a - h_{f1} - h_{f2} = H_b \tag{2.7}$$

In Equation (2.7), the head loss $(h_{fi})$ in the $i^{th}$ pipe is derived from either the Darcy–Weisbach or the Hazen–William formula. In either case, the head loss $(h_f)$ for a length of a pipe may be expressed in terms of the discharge in the pipe $(Q)$ as:

$$h_f = KQ^n \tag{2.8}$$

Considering the Darcy–Weisbach formula, the factor $K$ may be expressed as follows:

$$K = \frac{fL}{DA^2(2g)} \tag{2.9}$$

In Equation (2.9), $f$ is the friction coefficient, and $L$, $D$, and $A$ are the length, diameter, and cross-sectional area of the pipe, respectively. The term $g$ in the denominator is the acceleration due to gravity. Denoting the constant for the pipe $i$ as $K_i$ and discharge as $Q_i$, Equation (2.7) provides the first energy balance equation as under:

$$K_1 Q_1^2 + K_2 Q_2^2 = H_a - H_b \tag{2.10}$$

Similarly, another equation from energy considerations may be written between tanks $a$ and $c$ as below:

$$K_1 Q_1^2 + K_3 Q_3^2 = H_a - H_c \tag{2.11}$$

Equations (2.6), (2.10), and (2.11) form a set of three equations in terms of the three unknown discharges $Q_1$, $Q_2$, and $Q_3$. Note that the unknown pressure head at the lone junction, though an unknown, does not appear in the system of equations. Further, the continuity equation, Equation (2.6), is the only linear equation, while the two energy equations, Equation (2.10) and (2.11), are both non-linear since they contain higher powers of the unknown discharge variables. Nevertheless, the three equations may be expressed in terms of the three unknowns in a matrix form as under:

$$\begin{bmatrix} 1 & -1 & -1 \\ K_1 Q_1 & K_2 Q_2 & 0 \\ K_1 Q_1 & 0 & K_3 Q_3 \end{bmatrix} \begin{Bmatrix} Q_1 \\ Q_2 \\ Q_3 \end{Bmatrix} = \begin{Bmatrix} 0 \\ H_a - H_b \\ H_a - H_c \end{Bmatrix} \tag{2.12}$$

Equation (2.12) forms a system of simultaneous non-linear equations, which have to be solved either iteratively, or by implementing calculus, as explained in the

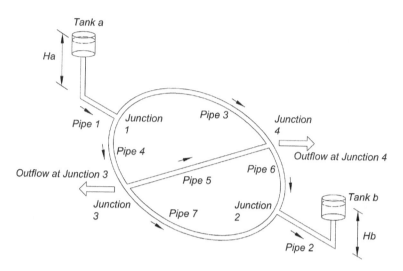

**FIGURE 2.6**   Flow distribution in a network of seven pipes connecting two reservoirs.

following section. However, the above method is now applied to a slightly more complex network as shown in Figure 2.6.

In the network shown in Figure 2.6, there are four junctions and, thus, four continuity equations may be formed, one for each, in terms of the discharges of the respective pipes meeting at a junction. Similarly, there are two closed loops and two energy balance equations, which may be written corresponding to each loop. Finally, there are two overhead tanks and thus one independent equation may be written for the "open" loop connecting the two reservoirs. The seven equations so generated may be expressed in a matrix form, and in terms of the unknown discharges of the seven pipes, designated by $Q_1$ through $Q_7$ (Equation 2.13). The hydraulic grade line connecting the known or unknown pressure heads, as was shown in Figure 2.5, have not been indicated in Figure 2.6 since the information is not required for the computation of the discharge through the pipes. However, if the unknown heads at the junctions are required, they may be obtained from the known pressure heads at the reservoirs and the pipe discharges, once they have been solved from Equation (2.13).

$$
\begin{bmatrix}
1 & 0 & -1 & -1 & 0 & 0 & 0 \\
0 & 0 & 1 & 0 & 1 & -1 & 0 \\
0 & 0 & 1 & 0 & -1 & 0 & -1 \\
0 & -1 & 0 & 0 & 0 & 1 & 1 \\
0 & 0 & K_3Q_3 & -K_4Q_4 & -K_5Q_5 & 0 & 0 \\
0 & 0 & 0 & 0 & K_5Q_5 & K_6Q_6 & -K_7Q_7 \\
K_1Q_1 & K_2Q_2 & K_3Q_3 & 0 & 0 & 0 & K_7Q_7
\end{bmatrix}
\begin{bmatrix}
Q_1 \\ Q_2 \\ Q_3 \\ Q_4 \\ Q_5 \\ Q_6 \\ Q_7
\end{bmatrix}
=
\begin{Bmatrix}
0 \\ 0 \\ 0 \\ 0 \\ 0 \\ 0 \\ H_a - H_b
\end{Bmatrix}
\qquad (2.13)
$$

The first four rows of the system of equations in Equation (2.13) represent the four linear equations for discharge continuity at the junctions, while the lower three are the non-linear energy balance equations – two around the two closed loops and one along the "open" loop from reservoir $a$ to reservoir $b$. The flow directions are arbitrarily chosen. If the given conditions dictate a reverse flow in any pipe, it would automatically be computed while solving the system of equations. The combined system of equations is non-linear, and the technique of solving the equations simultaneously is presented in one of the subsequent sections.

### 2.2.3　Derivation of the Unit Hydrograph

The Unit Hydrograph (UH) concept, first proposed by Sherman (Sherman, 1932), is still used occasionally as an useful tool in hydrology for estimating the runoff from a catchment with given rainfall inputs. The determination of the UH ordinates for a catchment from observed rainfall rates with time and concurrent stream-flow data requires some mathematical manipulations, and several methods have been proposed. For computer calculations, however, methods such as the least-square linear regression or optimization methods like the linear-programming technique are more convenient (Chow et al., 1988). Either method is facilitated by first expressing the time-sequenced effective rainfall and catchment outflow discharge ordinates in a matrix form. Denoting the number of rainfall ordinates by M, and the number of known discharge values by $N$, the number of UH ordinates is calculated as $N - M + 1$. In the example shown below, with $M = 3$, that is, 3 rainfall ordinates, $P_1$, $P_2$, and $P_3$, and $N = 6$, with 6 discharge ordinates, $Q_1$, $Q_2$, $Q_3$, $Q_4$, $Q_5$, and $Q_6$, the equivalent matrix relation turns out to be as follows:

$$
\begin{bmatrix}
P_1 & 0 & 0 & 0 \\
P_2 & P_1 & 0 & 0 \\
P_3 & P_2 & P_1 & 0 \\
0 & P_3 & P_2 & P_1 \\
0 & 0 & P_3 & P_2 \\
0 & 0 & 0 & P_3
\end{bmatrix}
\begin{Bmatrix}
U_1 \\ U_2 \\ U_3 \\ U_4
\end{Bmatrix}
=
\begin{Bmatrix}
Q_1 \\ Q_2 \\ Q_3 \\ Q_4 \\ Q_5 \\ Q_6
\end{Bmatrix}
\qquad (2.14)
$$

Note that the number of UH ordinates is found out as $6 - 3 + 1 = 4$. Thus,

$$[P][U] = [Q] \qquad (2.15)$$

The objective is to compute the ordinate values of the UH, that is, $U_1$, $U_2$, $U_3$, and $U_4$. However, the system of five equation represented by Equation (2.14) may not yield unique values for the three unknowns since there are more equations than the number of unknowns. Taking help of the linear regression technique, the best solution may be achieved that results in the least error, defined by a certain metric. We proceed by first defining a matrix $[Z]$, which is a product of the transpose of the matrix $[P]$ with itself, resulting in a square matrix, as defined below:

$$[Z] = [P]^T [P] \tag{2.16}$$

The vector of unknowns $[U]$ is then obtained by solving the set of equations obtained from Equation (2.15) as:

$$[Z][U] = [P]^T [Q] \tag{2.17}$$

The simultaneous solution of the set of linear Equation (2.17) may be obtained by standard methods, as discussed in the following section.

## 2.3   SOLUTION TECHNIQUES

For the non-linear and simultaneous equations discussed in Section 2.2, the solution techniques are briefly discussed in this section. The details are not greatly elaborated as much of these methods may be found in popular books devoted to numerical techniques, such as Chapra and Canale (2021) or Jain et al. (2019).

### 2.3.1   NON-LINEAR EQUATIONS IN ONE VARIABLE

Of the different numerical methods available for solving non-linear equations, the Newton–Raphson method (Chapra and Canale, 2021) is chosen for solving the problems in this section. For a function $f(x)$, written in terms of the independent variable $x$ and the derivative denoted by $f'(x)$, the following iterative steps may be followed to arrive at a converged value of the unknown within a specified error tolerance, starting from a trial solution $x_i$. Both the function and its derivative are evaluated at $x_i$, and an updated trial solution ($x_{i+1}$) is evaluated by the following algorithm:

$$x_{i+1} = x_i - \frac{f(x_i)}{f'(x_i)} \tag{2.18}$$

The examples on finding the normal depth of flow in a trapezoidal channel, or the depth of flow in a constricted and raised channel section illustrated in Section 2.1 fall under this category. Instead of solving each problem individually, we demonstrate the generic procedure that may be followed for solving the equations. A generic algorithm for the purpose is shown below and needs to be adapted for the specific problem.

```
define f(x)
define f'(x)
initiate values xstart, e_relative
xi = xstart
while abs((xi+1-xi)/xi) >e_relative
     xi+1 = xi -f(x)/ f'(x)
end while
```

The values of $x_{start}$ and e_relative are inputs to be defined by the user and the iterations are continued until the relative error reduces below a specified value of e_rela-
tive. Smaller the value of e_relative, more accurate would the solution be, but at the
cost of a greater number of iterations.

## 2.3.2 LINEAR SIMULTANEOUS EQUATIONS

Solving a set of linear equations simultaneously, among the examples presented in
Section 2.1, appears in the steady-state analysis of a system of interconnected reactors and in the derivation of the unit hydrograph.

The Gaussian-Elimination method may be used for such cases, the basic steps of
which are explained below for the simple elimination technique, without pivoting.
Extension of the algorithm to include "pivoting", which is required when some of the
diagonal elements are non-dominant, may be found in standard texts on numerical
methods like Chapra and Canale (2021). We may express the set of equations in the
following form:

$$a_{11}x_1 + a_{12}x_2 + a_{13}x_3 + \ldots a_{1n}x_n = b_1$$
$$a_{21}x_1 + a_{22}x_2 + a_{23}x_3 + \ldots a_{2n}x_n = b_2$$
$$a_{31}x_1 + a_{32}x_2 + a_{31}x_3 + \ldots a_{3n}x_n = b_3$$
$$.$$
$$.$$
$$a_{n1}x_1 + a_{n2}x_2 + a_{n1}x_3 + \ldots a_{nn}x_n = b_1$$

(2.19)

The algorithm for implementing the Gaussian-Elimination method involves two
parts, the first being a progressive row-by-row elimination of the variables, which
may be written as under:

```
do k from 1 to n-1
        do i = k+1, n
                p = ai,k / ak,k
                do j = k+1, n
                        ai,j = ai,j - p * ak,j
                end do
                bi = bi - p * bk
        end do
end do
```

The above step is followed by the back-substitution phase, the algorithm for
which may be framed as given below:

```
Xn = bn / an,n
do i from n-1 to n, in steps of -1
        sum = 0
        do j = i+1, n
                sum = sum  = ai,k * xj
```

```
        end do
        xi = (bi - sum) / ai,i
end do
```

### 2.3.3 NON-LINEAR SIMULTANEOUS EQUATIONS

Simultaneous equations containing non-linear terms may also be solved by extending the Newton–Raphson method accordingly (Kopchenova and Maron, 1981). The problem of flow distribution in pipe networks under a steady-state condition falls under this category. Here, some or all the coefficients $a_{ij}$ in the simultaneous equations are also functions of the variables $x_i$. In this case, we write the system of equations as a set of functions as given below:

$$f_1(x_1, x_2, \ldots, x_n) = 0 \tag{2.20}$$
$$f_2(x_1, x_2, \ldots, x_n) = 0$$

.
.
.
.

$$f_n(x_1, x_2, \ldots, x_n) = 0$$

The arguments of the functions, that is, the variables $x_1$, $x_2$, ..., $x_n$ may be considered as an $n$-dimensional vector $[x]^T = [x_1 \quad x_2 \quad \ldots \quad \ldots \quad x_n]^T$. The functions $f_1$, $f_2$, ..., $f_n$ may also be written as an $n$-dimensional vector function of the form $[f(x)]^T = [f_1 \quad f_2 \quad \ldots \quad \ldots \quad f_n]^T$. Thus, the set of equations represented by Equation (2.20) may be concisely expressed as:

$$f(x) = 0 \tag{2.21}$$

The system of Equations (2.21) may be solved by the method of successive approximation. If the $p^{th}$ approximation of the variables $[x]^{(p)T} = [x_1 \quad x_2 \quad \ldots \quad \ldots \quad x_n]^{(p)T}$ is evaluated or known, then the $(p+1)^{th}$ approximation may be found as:

$$x^{(p+1)} = x^{(p)} + \varepsilon^{(p)} \tag{2.22}$$

where $[\varepsilon]^{(p)T} = [\varepsilon_1 \quad \varepsilon_2 \quad \ldots \quad \ldots \quad \varepsilon_n]^{(p)T}$ is the vector of errors given by

$$\varepsilon^{(p)} = -J^{-1} f\left(x^{(p)}\right) \tag{2.23}$$

In Equation (2.23), $J$ is the Jacobian matrix as given below, and is evaluated for $x = x^{(p)}$.

$$J = \begin{bmatrix} \dfrac{\partial f_1}{\partial x_1} & \dfrac{\partial f_1}{\partial x_2} & \cdot\;\cdot & \cdot & \dfrac{\partial f_i}{\partial x_n} \\[2ex] \dfrac{\partial f_2}{\partial x_1} & \dfrac{\partial f_2}{\partial x_2} & \cdot\;\cdot & \cdot & \dfrac{\partial f_2}{\partial x_n} \\[2ex] \cdot & \cdot & & \cdot \\[1ex] \cdot & \cdot & & \cdot \\[1ex] \cdot & \cdot & & \cdot \\[1ex] \dfrac{\partial f_n}{\partial x_1} & \dfrac{\partial f_n}{\partial x_2} & \cdot\;\cdot & \cdot & \dfrac{\partial f_n}{\partial x_n} \end{bmatrix} \qquad (2.24)$$

The variables may be updated by the following formula:

$$x^{(p+1)} = x^{(p)} + \varepsilon^{(p)} \qquad (2.25)$$

The iterations are repeated till the desired accuracy is achieved. Solution of Equation (2.25) may be found by solving the equivalent linear set of equations as given below:

$$J\left(x^{(p)}\right)\varepsilon^{(p)} = -f\left(x^{(p)}\right) \qquad (2.26)$$

The algorithm for this step is similar to that of the solution for linear simultaneous equations presented in the previous section.

## 2.4   PYTHON PROGRAMS

This section demonstrates the algorithms discussed above by converting these into Python codes. These programs may be used as examples by the reader in writing similar, and perhaps more refined, codes for other problems in fluid flow and contaminant transport. Specifically, the codes demonstrate the numerical solution of non-linear equations in one variable and simultaneous solution of a system of linear and non-linear equations.

### 2.4.1   NON-LINEAR EQUATIONS IN ONE VARIABLE: FINDING UNIFORM FLOW DEPTH IN A CHANNEL

This problem seeks to find the normal depth ($y_n$) of flow in a trapezoidal channel from Equation (2.1), by modifying it as shown below:

$$nQ - AR^{2/3}S_0^{1/2} = 0 \qquad (2.27)$$

The steady-state discharge through the channel ($Q$), Manning's roughness coefficient ($n$), and the longitudinal slope of the channel ($S_0$) are considered given. The cross-sectional area ($A$) and wetted perimeter ($R$) contain the unknown variable $y_n$. In order to solve Equation (2.27), it is first written in the form of a function of the variable $y_n$ as follows (Chaudhry, 2008):

$$F\left(y_n\right) = A^{5/3} P^{-2/3} - \frac{nQ}{S_0^{1/2}} \tag{2.28}$$

The derivative of Equation (2.28) is found out next, as given below, which is required for use in the Newton–Raphson method. Note that the second term in Equation (2.28) is constant, depending only upon the specified numeric values of the data, and hence does not have a derivative.

$$\frac{dF\left(y_n\right)}{dy_n} = \frac{d}{dy_n}\left( A^{\frac{5}{3}} P^{-\frac{2}{3}} - \frac{nQ}{S_0^{\frac{1}{2}}} \right) = \frac{5}{3} P^{-2/3} A^{2/3} \frac{dA}{dy_n} - \frac{2}{3} A^{5/3} P^{-5/3} \frac{dP}{dy_n}$$
$$= \frac{5}{3} BR^{2/3} - \frac{2}{3} R^{5/3} \frac{dP}{dy_n} \tag{2.29}$$

Note that $\dfrac{dA}{dy_n} = B$ and $\dfrac{dP}{dy_n} = 2\sqrt{1+s^2}$. The following Python code shows the steps for computing the value of $y_n$ iteratively using the Newton–Raphson method. The variables used in this code are indicated in the following table.

| Variable | Description | Variable | Description |
|---|---|---|---|
| b0 | Base width of trapezoidal channel (m) | Q | Discharge (m³/s) |
| s | Side slope of trapezoidal channel (-) | yinitial | Initial guess of normal depth (m) |
| s0 | Longitudinal slope of channel (-) | errorallow | Allowable error tolerance for computed solution (m) |
| mn | Manning's roughness coefficient (s/m^{1/3}) | | |

```
# Normal depth of a trapezoidal channel using the Newton-
# Raphson method
import numpy as np
import matplotlib.pylab as plt
b0 = 20.0
s = 2.0
s0 = 0.001
mn = 0.020
Q = 40
```

```
yinitial = 1.0
errorallow = 0.0001

def area(y):
    area=y*(b0+s*y)
    return area

def wetperi(depth):
    wetperi = b0+2*depth*np.sqrt(1+s*s)
    return wetperi

def hyrad(depth):
    hyrad = (b0+depth*s)*depth/(b0+2*depth*np.sqrt(1+s*s))
    return hyrad

def topwidth(depth):
    topwidth=b0+2*s*depth
    return topwidth

iter = 0
max_error = 1.0
yn = yinitial
while (max_error > errorallow):
    iter = iter+1
    Fyn = area(yn)**1.6667/wetperi(yn)**0.6667-mn*Q/
np.sqrt(s0)
    dFdyn = (5/3)*topwidth(yn)*hyrad(yn)**0.6667\
    -(2/3)*hyrad(yn)**1.6667*(2+np.sqrt(1+s*s))
    ynew = yn - Fyn/dFdyn
    max_error = np.max(abs(ynew-yn))
    yn = ynew

print('iter = ',iter,'  Normal Depth = %.3f'%yn)
ymin = 0
ymax = 3
npoints = 100
y = np.linspace(ymin, ymax, npoints)
dy = (ymax-ymin)/npoints
fy = np.ones(npoints)
fy0 = np.zeros(npoints)

yy = ymin
for i in range(0,npoints):
    yy = ymin + float(i)*dy
    fy[i] = area(yy)**1.6667/wetperi(yy)**0.6667-mn*Q/
np.sqrt(s0)

fig = plt.figure()
ax = fig.add_subplot(1, 1, 1)
ax.plot(y, fy)
ax.plot(y, fy0)
```

```
ax.set_xlabel(' yn (m)')
ax.set_ylabel(' f(yn) ')

plt.show()
```

For given data values as mentioned in lines 5 to 11 of the code, the output obtained on running the code is as follows:

```
iter =  3    Normal Depth = 1.132
```

The normal depth function (Equation 2.28) is plotted in Figure 2.7 which shows that it becomes zero (that is where it intersects the horizontal line) at the obtained value of 1.132m.

### 2.4.2  NON-LINEAR EQUATIONS IN ONE VARIABLE: FINDING THE HEIGHT AND VELOCITY OF A SURGE WAVE

Here, we find the height and velocity of a surge wave that is produced when the initial steady-state flow condition of a channel is changed. We assume an initial depth and velocity of flow ($y_1$ and $v_1$, respectively) in a rectangular channel (of width B) that correspond to a discharge of $Q_1$. If the discharge is now suddenly increased to say, a value of $Q_2$, then we may calculate the corresponding increased depth of flow using the following formula (see Equation 2.3):

$$\left(v_1 - v_2\right)^2 = \frac{g(y_1 - y_{2)}}{2y_1 y_2}\left(y_1^2 - y_2^2\right)$$

(2.30)

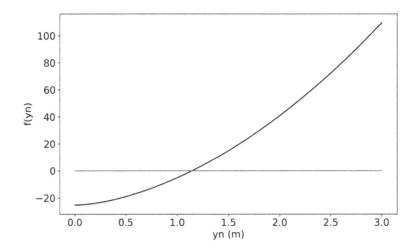

**FIGURE 2.7**   Variation of the non-linear function (Equation 2.28) in terms of normal depth, $y_n$. Solution for $y_n$ lies where the function's graph intersects the zero-value line (the *x*-axis). (Color image available in eBook).

We transform Equation (2.30) by substituting the following variables: (a) $q_1 = Q_1/B$; (b) $q_2 = Q_2/B$; (c) $v_1 = q_1/y_1$, thus leading to the following equation:

$$\left(v_1 y_2 - q_2\right)^2 - \frac{g}{2y_1} y_2(y_2 - y_1)\left(y_2^2 - y_1^2\right) = 0 \tag{2.31}$$

In Equation (2.31), the only unknown is $y_2$, which is solved using the Newton–Raphson method and implemented in the Python code given below. The variables used in this code are given in the following table.

| Variable | Description | Variable | Description |
|---|---|---|---|
| g | Acceleration due to gravity (m/s²) | errorallow | Allowable error tolerance for computed solution (m) |
| Q1 | Initial discharge (m³/s) | b0 | Width of rectangular channel (m) |
| Q2 | Increased discharge (m³/s) | | |

```
# Surge height and velocity by Newtion-Raphson method
import numpy as np
import matplotlib.pylab as plt

g = 9.81
Q1 = 5
Q2 = 10
y1 = 2.0
b0 = 2.5
errorallow = 0.0001
maxiter = 100
max_error = 1.0

q2 = Q2/b0
q1 = Q1/b0
v1 = q1/y1
k = g/(2*y1)

y2 = 4.0
iter = 0
while (max_error > errorallow):
    iter = iter+1
    if(iter > maxiter): break
    fy2 = (v1*y2-q2)**2-k*y2*(y2-y1)*(y2**2-y1**2)
    dfdy2 = 2*(y2-q2)*v1-\
    k*(2*y2**2*(y2-y1)+y2*(y2**2-y1**2)+(y2-y1)*(y2**2-y1**2))
    y2 -= fy2/dfdy2
    max_error = abs(fy2/dfdy2)
```

```
v2 = q2/y2
vw = (v2*y2-v1*y1)/(y2-y1)
print('Surge height (y2)= %.3f'%y2,'m and Wave velocity (vw)=
%.3f'%vw,'m/s')

ymin = 0
ymax = 3
npoints = 100
y = np.linspace(ymin, ymax, npoints)
dy = (ymax-ymin)/npoints
fy = np.ones(npoints)
fy0 = np.zeros(npoints)

yy = ymin
for i in range(0,npoints):
    yy = ymin + float(i)*dy
    fy[i] = (v1*yy-q2)**2-k*yy*(yy-y1)*(yy**2-y1**2)

fig = plt.figure()
ax = fig.add_subplot(1, 1, 1)
ax.plot(y, fy)
ax.plot(y, fy0)
ax.set_xlabel(' y2 (m)')
ax.set_ylabel(' f(y2) ')

plt.show()
```

For the given data values as mentioned in lines 5 to 10 of the code, the output obtained by running the code is:

```
Surge height (y2)= 2.334 m and Wave velocity (vw)= 5.981 m/s
```

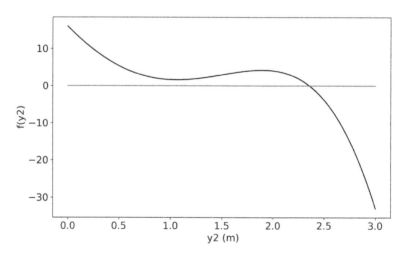

**FIGURE 2.8** Variation of the non-linear function (left side of Equation 2.31) in terms of surge height, $y_2$. Solution for $y_2$ is where the function's value intersects the zero-value line (the $x$-axis). (Color image available in eBook).

The non-linear function involving $y_2$ (Equation 2.31) is plotted in Figure 2.8 to check where it becomes zero (that is where it intersects the horizontal zero-value line), which is seen to be at the obtained value of 2.334 m.

### 2.4.3 NON-LINEAR EQUATIONS IN ONE VARIABLE: FINDING THE DEPTH OF FLOW ABOVE A HUMP IN A CONTRACTION

The third problem involves solving the depth of flow at a contracted and humped portion of a rectangular channel. The specific energies on the upstream and in the section of the channel under consideration relate the two depths $y_1$ and $y_2$ in the respective sections by the following non-linear equation in terms of the flow depth $y_2$ (see Equation 2.4):

$$y_1 + \frac{q_1^2}{2\,gy_1^2} = y_2 + \frac{q_2^2}{2\,gy_2^2} + \Delta z \qquad (2.32)$$

It is assumed that the total discharge $Q$ and the depth of flow in the wider section, $y_1$, are known. Further, the widths $b_1$ and $b_2$ at the two sections, corresponding to the normal and contracted sections, respectively, are also assumed to be known along with the height of the rise, $\Delta z$. The equation is solved for the unknown depth $y_2$ by first rewriting Equation (2.32) as a function of the variable as under:

$$y_2^3 - y_2^2\left[ y_1 + \frac{q_1^2}{2\,gy_1^2} - \Delta z \right] - \frac{q_2^2}{2\,g} = 0 \qquad (2.33)$$

Since the function on the left of the equality in Equation (2.33) is non-linear in terms of the variable $y_2$, the method of Newton–Raphson is used in the following Python code for finding its solution. Please note that the derivative of the function on the left of Equation (2.33) in terms of $y_2$ is evaluated and used in the code. The variables used in this code are given in the following table. Note that height of the hump, $\Delta z$, is an input variable in the code, which is used in checking the variations of flow depth and velocity at the contracted section as the height of the hump is changed.

| Variable | Description | Variable | Description |
| --- | --- | --- | --- |
| g | Acceleration due to gravity (m/s²) | b1 | Width of channel at upstream section (m) |
| Q | Discharge (m³/s) | b2 | Width of channel at contracted section (m) |
| y1 | Depth of flow at upstream (non-contracted) section (m) | errorallow | Allowable error tolerance for computed solution (m) |
| Dz[i] | The height of the hump (m) is input as an array | | |

```
# Depth of flow over a hump with a contraction  -  with hump
height varied
import numpy as np
import matplotlib.pylab as plt
```

```
g = 9.81
Q = 10
y1 = 2.0
b1 = 4.0
b2 = 3.5
errorallow = 0.0001
maxiter = 100
max_error = 1.0

q1 = Q/b1
q2 = Q/b2
v1 = q1/y1
k1 = q1**2/(2*g)
k2 = q2**2/(2*g)

dzmin = 0
dzmax = 0.675
npoints = 100
dx = (dzmax-dzmin)/npoints
yy = np.zeros(npoints)
vv = np.zeros(npoints)
Dz = np.zeros(npoints)

for i in range(0,npoints):
    Dz[i] = dzmin + float(i)*dx
    y2 = 4.0
    iter = 0
    max_error = 1.0
    while (max_error > errorallow):
        iter = iter+1
        if(iter > maxiter): break
        fy2 = y2**3-y2**2*(y1+k1/y1**2-Dz[i])+k2
        dfdy2 = 3*y2**2-2*y2*(y1+k1/y1**2-Dz[i])
        y2 -= fy2/dfdy2
        max_error = abs(fy2/dfdy2)
    yy[i] = y2
    vv[i] = q2/y2

fig = plt.figure()
ax = fig.add_subplot(1, 1, 1)
ax.plot(y, yy, label = 'Depth over hump (m)')
ax.plot(y, vv, label = 'Velocity over hump (ms-1)')
ax.set_xlabel(' Height of hump (m)')
ax.set_ylabel(' Depth (m) and velocity (m/s) over hump ')
leg = ax.legend()
plt.show()
```

On running the code, the corresponding graphical output (Figure 2.9) is obtained which shows that for a value of the hump height ($\Delta z$) greater than about 0.675m, Equation (2.33) does not yield correct results for the computed depth of flow. This is the condition of flow "choking", whereby the flow depth and velocity reach the "critical" condition and it is not possible for the flow depth to reduce any further.

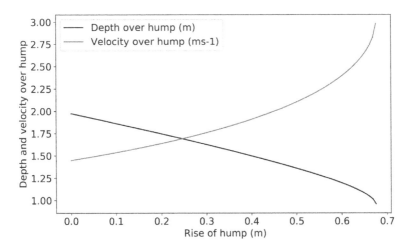

**FIGURE 2.9** Variation of depth and flow velocity over a hump (located at the contraction of a channel), as a function of the hump height. (Color image available in eBook).

A calculation shows that for the data given in lines 5 to 12 of the code, a depth of about 0.832 m above the hump produces the critical flow (that is, when Froude number = 1.0).

### 2.4.4 SOLUTION OF LINEAR SIMULTANEOUS EQUATIONS: CONCENTRATIONS IN INTERCONNECTED REACTORS

The system of interconnected reactors with steady incoming and outgoing flows, carrying a conservative contaminant dissolved with the flow, (Figure 2.4) is analysed, which represents a typical case of a solution of linear simultaneous equations. The system of equations developed (Equation 2.5) is now recast in the form $[A]\{x\} = \{b\}$, as a first step before converting it into a code.

$$\begin{bmatrix} Q_{12}+Q_{15} & 0 & -Q_{31} & 0 & 0 \\ -Q_{12} & Q_{25}+Q_{24} & Q_{23} & 0 & 0 \\ 0 & -Q_{23} & Q_{31}+Q_{34} & 0 & 0 \\ 0 & Q_{24} & Q_{34} & -Q_{40} & Q_{54} \\ Q_{15} & Q_{25} & 0 & 0 & -Q_{54}-Q_{50} \end{bmatrix} \begin{Bmatrix} c_1 \\ c_2 \\ c_3 \\ c_4 \\ c_5 \end{Bmatrix} = \begin{Bmatrix} Q_{01} \cdot c_{01} \\ 0 \\ Q_{03} \cdot c_{03} \\ 0 \\ 0 \end{Bmatrix} \quad (2.34)$$

Following the notations introduced previously, $Q_{ij}$ represent the discharge from tank $i$ to tank $j$, while $c_i$ stands for concentration in the $i^{th}$ tank. The notation $c_{0i}$ represents the concentration of liquid entering tank $i$, along with the flow $Q_{0i}$, while $Q_{i0}$ stands for the flow leaving the system from tank $i$.

The equivalent Python code is as follows, which is self-explanatory:

```
# Simultaneous solution of concentrations in reactor system
import numpy as np
```

```python
import matplotlib.pylab as plt

Q01 = 10; Q03 = 2 # Inflows
Q40 = 6; Q50 = 6 # Outflows
Q12 = 7; Q15 = 5
Q23 = 1; Q24 = 2; Q25 = 4
Q31 = 2; Q34 = 1
Q40 = 6
Q54 = 3;
c01 = 0.1; c03 = 0.2 # Concentrations of incoming flows

A = np.array([[Q12+Q15,0,-Q31,0,0],
              [-Q12,Q25+Q24,Q23,0,0],
              [0,-Q23,Q31+Q34,0,0],
              [0,Q24,Q34,-Q40,Q54],
              [Q15,Q25,0,0,-Q54-Q50]])
b = np.array([Q01*c01,0,Q03*c03,0,0])
c = np.linalg.solve(A,b)

print("Concentrations in reactors = ",c)
cmin = 0
cmax = 1
npoints = 10
c1 = np.linspace(cmin, cmax, npoints)
c2 = np.zeros(npoints)
c4 = np.zeros(npoints)
dc = (cmax-cmin)/npoints

cc = cmin
for i in range(0,npoints):
    cc = cmin + float(i)*dc
    c01 = cc
    b = np.array([Q01*c01,0,Q03*c03,0,0])
    c = np.linalg.solve(A,b)
    c2[i] = c[2]
    c4[i] = c[4]

fig = plt.figure()
ax = fig.add_subplot(1, 1, 1)
ax.plot(c1,c3, label='c3')
ax.plot(c1,c5, label='c5')
ax.set_xlabel(' c01 ')
ax.set_ylabel(' c3 and c5 ')
leg = ax.legend()
plt.show()
```

The output obtained from running the code with the data given in lines 5 to 12 is a statement showing the concentrations in the different reactors and a graph (Figure 2.10) that depicts the variations in the concentrations of tanks 3 and 5 of Figure 2.4, as a function of the incoming concentration $c_{0i}$. The plot shows that the concentrations in the two reactors do not increase by the same amount, which may be judged

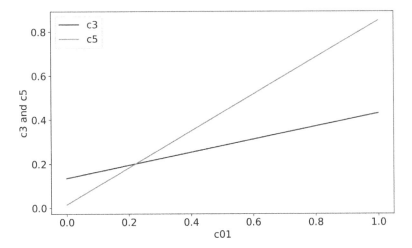

**FIGURE 2.10**  Variation of the contaminant concentrations c3 and c5, in tanks 3 and 5, respectively, as a function of the concentration of the incoming concentration c01. (Color image available in eBook).

easily by following the configuration of the interconnected piping systems and the assumed flows between the tanks. Interestingly, the concentration of tank 3 (blue line in Figure 2.10) is lower than the concentration of tank 5 (orange line in Figure 2.10) for $c_{0i}$ of about 0.25, beyond which the concentration of tank 5 exceeds that of the other.

```
Concentrations in reactors = [0.11121495 0.10186916 0.16728972
0.11536864 0.10706127]
```

### 2.4.5  SOLUTION OF LINEAR SIMULTANEOUS EQUATIONS: DERIVATION OF THE UNIT HYDROGRAPH

This example is taken from hydrology and it demonstrates several functions from linear algebra in Python. We first assume a unit hydrograph and a precipitation sequence, and obtain the resulting direct runoff hydrograph. We may call this step as the forward problem. Next, we use the generated direct runoff hydrograph and the assumed precipitation and find out the unit hydrograph by an inverse problem, the steps of which are discussed in Equations (2.14–2.17).

The Python code for the forward problem is given below:

```
# Finding the direct runoff hydrograph from given Unit
# Hydrograph and precipitation
import numpy as np

UH = np.array([5,10,20,17,12,8,5,2])
precipitation = np.array([2,6,4])
```

```
M = len(precipitation)
K = len(UH)
N = K+M-1
P = np.zeros((N,K))

for j in range(0,K):
    for i in range(0,K):
        for l in range(0,M):
            row = j+l
            P[row,j] = precipitation[l]

Q = np.dot(P,UH)
print(Q)
```

The variables used in the above program are as follows:

| Variable | Description | Variable | Description |
|---|---|---|---|
| UH | An array containing $K$ unit hydrograph ordinates. In the data given, $K$ = 8 | precipitation | An array containing $M$ precipitation ordinates. In the data given, $M$ = 3 |
| Q | An array containing $N$ direct runoff hydrograph ordinates. Here, $N$ = 8 + 3 −1 = 10 | P | The array of arranged precipitations in matrix form in Equation (2.14) |

The unit hydrograph and direct runoff hydrograph ordinates are both in consistent units $[M^3T^{-1}]$. The precipitation ordinates are in units of depths [L] and defined according to the problem given. Since the above program is only demonstrative, the duration or dimension of the hydrographs are not mentioned and may be set according to the given problem.

On running the above code, the following output is obtained:

```
Direct runoff hydrograph ordinates are: [10. 50. 120.194.
206.156. 106.66. 32. 8.]
```

For the inverse problem, the unit hydrograph ordinates are derived from the direct runoff hydrograph by making use of the following program in Python. The direct runoff hydrograph ordinates used as input to the code are those obtained from the forward problem.

```
# Unit hydrograph derivation by matrix calculations
import numpy as np
import matplotlib.pylab as plt

precipitation = np.array([2,6,4])
Q = [10.,50.,120.,194.,206.,156.,106. ,66. ,32.,8.]
```

```
M = len(precipitation)
N = len(Q)
K = N-M+1

P = np.zeros((N,K))
for j in range(0,K):
    for i in range(0,K):
        for l in range(0,M):
            row = j+l
            P[row,j] = precipitation[l]

Ptrans = np.zeros((K,N))
for i in range(0,N):
    for j in range(0,K):
        Ptrans[j,i] = P[i,j]

Z = np.dot(Ptrans,P)
rhs = np.dot(Ptrans,Q)
UH = np.linalg.solve(Z,rhs)
print ("Unit hydrograph ordinates:",UH)

UH1 = np.zeros(N)
for i in range(0,K):
    UH1[i]=UH[i]
time = np.linspace(0, N, N)
fig = plt.figure()
ax = fig.add_subplot(1, 1, 1)
plt.plot(time, Q, label='Direct Runoff Hydrograph (DRH)')
plt.plot(time, UH1, label='Unit Hydrograph (UH)')
leg = ax.legend()
ax.set_xlabel(' Time ')
ax.set_ylabel(' DRH and UH ')

plt.show()
```

The variables used in the above program are similar to those of the code written for the forward problem. Some other variables which are unique to this code are described below:

| Variable | Description | Variable | Description |
| --- | --- | --- | --- |
| Ptrans | Transpose of matrix [P], that is, $[P]^T$ | rhs | The right hand side vector of Equation (2.17) |
| Z | The product $[P]^T[P]$ | UH | The unit hydrograph ordinates obtained by solving Equation (2.17) |

The output of the above Python code prints the unit hydrograph ordinates as shown below and then plots both the direct runoff as well as the unit hydrographs on a common graph as in Figure 2.11.

```
Unit hydrograph ordinates: [5. 10.20. 17.12. 8.5. 2.0. 0.]
```

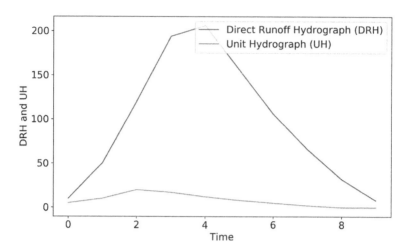

**FIGURE 2.11** Plot of the given Direct Runoff Hydrograph (DRH, in blue) and the derived Unit Hydrograph (UH, in orange), against time. (Consistent units may be followed). (Color image available in eBook).

### 2.4.6 SOLUTION OF NON-LINEAR SIMULTANEOUS EQUATIONS: FLOW DISTRIBUTION IN A THREE-PIPE NETWORK

The flow distribution in the different branches of a pipe network demonstrates an example of a system of simultaneous non-linear equations. The flows in the networks shown in Figure 2.5 and 2.6 are expressed by Equations (2.12) and (2.13), respectively, and the method of Newton–Raphson may be applied to solve these, as shown below.

Choosing Equation (2.12), we first evaluate the Jacobian matrix using initially guessed values of discharges in each pipe, and then solve the following system of equations to compute the array of discharge errors, $\{\Delta Q\}$.

$$\begin{bmatrix} 1 & -1 & -1 \\ 2K_1Q_1 & 2K_2Q_2 & 0 \\ 2K_1Q_1 & 0 & 2K_3Q_3 \end{bmatrix} \begin{bmatrix} \Delta Q_1 \\ \Delta Q_2 \\ Q\Delta_3 \end{bmatrix} = \begin{Bmatrix} Q_1 - Q_2 - Q_3 \\ K_1Q_1^2 + K_1Q_1^2 - H_a + H_b \\ K_1Q_1^2 + K_3Q_3^2 - H_a + H_c \end{Bmatrix} \qquad (2.35)$$

Once found, the updated values of discharges are estimated as follows:

$$\begin{Bmatrix} Q_1 \\ Q_2 \\ Q_3 \end{Bmatrix}^{updated} = \begin{Bmatrix} Q_1 \\ Q_2 \\ Q_3 \end{Bmatrix}^{old} - \begin{Bmatrix} \Delta Q_1 \\ \Delta Q_2 \\ \Delta Q_3 \end{Bmatrix} \qquad (2.36)$$

The process is repeated by replacing the old values of discharges with those updated, till the maximum absolute value of the errors drops below a specified tolerance level. The following Python program demonstrates these steps for the three-reservoir problem of pipe network (Figure 2.5).

```
# 3-pipe network with 3 tanks
import numpy as np

g = 9.81
errorallow = 0.0001
maxiter = 10
nQ = 3
Q = np.array([0.4,0.2,0.2])

njunctions = 1
junctionQ = np.array([+1,-1,-1])

nloops = 2
loopQ = np.array([[+1,+1,0],
                  [+1,0,+1]])
loopH =  np.array([[100,70],
                   [100,30]])

length = np.array([1000,600,500])
diameter = np.array([0.254,0.305,0.152])
friction = np.array([0.020,0.018,0.025])
area = 3.14*diameter**2/4
k = friction*length/(2*g*diameter*area**2)

iter = 0
max_error = 1.0
while (max_error > errorallow):
    iter = iter+1
    A = np.zeros((nQ,nQ))
    rhs = np.zeros(nQ)
    if(iter > maxiter): break

    for i in range(0,njunctions):
        for j in range(0,nQ):
            A[i,j] = junctionQ[j]
            rhs[i] += junctionQ[j]*Q[j]

    for i in range(0,nloops):
        for j in range(0,nQ):
            A[i+njunctions,j] = loopQ[i,j]*k[j]*Q[j]*2
            rhs[i+njunctions] += loopQ[i,j]*k[j]*Q[j]**2
        rhs[i+njunctions] -= (loopH[i,0]-loopH[i,1])

    delQ = np.linalg.solve(A,rhs)
    Q -= delQ
    max_error = np.max(abs(delQ/Q))

print("iter = ",iter,". Discharges in the pipes are as
follows:")
print(Q)
```

The variables used in this program are described below:

| Variable | Description | Variable | Description |
|---|---|---|---|
| g | Acceleration due to gravity (m/s²) | length | Array storing the lengths of the pipes |
| nQ | The number of unknown discharges (or pipes) | diameter | Array storing the diameters of the pipes |
| Q | Array of initial guessed value of discharges in the pipes (m³/s) | friction | Array storing the friction factors of the pipes |
| njunctions | The number of pipe junctions | area | Array of computed areas of the pipes |
| junctionQ | Array indicating pipe connection at junctions | k | Constant in the frictional head formula for a pipe $h_f = kQ^2$ |
| nloops | Number of energy balance loops (closed or open) | | |
| loopQ | Array indicating pipes around or along a loop | maxiter | Maximum permitted iterations |
| loopH | Array for starting and closing pressure heads around or along a loop | errorallow | Allowable error tolerance for computed solution (m) |

The following details may be noted for the input data arrays: junctionQ, loopQ, and loopH.

1. junctionQ: The number of rows of this array corresponds to the number of junctions, while the number of columns stands for the number of pipes in the system. Since there is only one junction node and three branches of pipes in the system considered, the junctionQ array here is a single-rowed, three-columned array. Each element of this array is represented by either a +1 or −1, indicating whether the corresponding pipe is flowing into or out of the junction. For example, in the above code, the junctionQ array values are input as ( [+1,-1,-1] ), meaning that pipe 1 is entering the junction, while the other two are leaving the junction. For more branches in the system, there may be some branches which are not connected to the system, as shown in another example.

2. loopQ: This is a matrix with each row corresponding to one loop of the system, whether open or closed. In the present problem, there is no closed loop, but because of the three overhead water tanks, there are two open loops that may be considered to originate at a higher potential point and ending up at a lower. Each loop has two pipes connecting the fixed head points and this is reflected by the values provided in each row of the matrix.

For example, in the first row, which corresponds to the open loop running from tank *a* to tank *b*, pipes 1 and 2 are the joining conduits between the two and are both given the code +1, indicating the assumed positive direction of flow in these pipes. The third element of the first row of the matrix is given a value of zero, suggesting that pipe 3 is not a part of the first open loop. The second row of the matrix, similarly, stands for the second loop assumed between tanks *a* and *c*, and since only pipes 1 and 3 appear along this path and, that too, in the assumed direction of flows, the elements have the values [+1,0,+1].

3. loopH: This matrix, having rows equal to the number of loops, but with only two columns, is used to input data about the known pressure heads at the two ends of an open loop. For example, the array values in the first row are [100,70], meaning that the pressure head at the starting of the loop is 100 m, while that at the farthest end is 70m. Although there is no closed loop in the particular example considered, the array may be modified for systems with closed loops by entering an arbitrary pair of equal values (which may as well be given as zero-s), since the energy balance equation for such a loop starts and ends at the same point. This is demonstrated for a more complicated network of pipes in the following section.

Note that in the program, the final solution is obtained iteratively with the computations carried on until a given error tolerance criteria is achieved. When run, the program gives the following output:

```
iter = 6 . Discharges in the pipes (in m3/s) are as follows:
[0.13382734 0.07643163 0.05739571]
```

### 2.4.7  SOLUTION OF NON-LINEAR SIMULTANEOUS EQUATIONS: FLOW DISTRIBUTION IN A GENERAL PIPE NETWORK

In this section, the Python code for solving a general pipe network is presented. The seven-branched network shown in Figure 2.6 is considered and a Python code, with slight modification to that of the preceding example, is used. In this case, there are two closed loops, apart from a single open loop, obtained by balancing energy between the two reservoirs of the system. Further, there are four junctions in the system which permit us to write four continuity equations. The program also includes the data of outflows, or withdrawals, from junctions.

Following the assumed flow directions as shown in Figure 2.6, the program in Python accepts data in a similar fashion as the one demonstrated in the previous example.

```
# 7-pipe network with 2 tanks. Outflow from junctions
# specified.
import numpy as np

g = 9.81
errorallow = 0.0001
maxiter = 10
```

```
nQ = 7
Q = np.array([0.2,0.2,0.2,0.2,0.2,0.2,0.2])
njunctions = 4
junctionQ = np.array([[+1,0,-1,-1,0,0,0],
                       [0,-1,0,0,0,+1,+1],
                       [0,0,0,+1,-1,0,-1],
                       [0,0,+1,0, 1,-1,0]])

junctionO = np.array([0.0,0.0,0.1,0.05])

nloops = 3
loopQ = np.array([[0,0,+1,-1,-1,0,0],
                  [0,0,0,0,+1,+1,-1],
                  [+1,+1,+1,0,0,+1,0]])
loopH =  np.array([[0,0],
                   [0,0],
                   [100,50]])

length = np.array([2000,2000,1000,500,1000,500,1000])
diameter = np.array([0.50,0.50,0.40,0.40,0.40,0.40,0.40])
friction = np.array([0.02,0.02,0.02,0.02,0.02,0.02,0.02])
area = 3.14*diameter**2/4
k = friction*length/(2*g*diameter*area**2)

iter = 0
max_error = 1.0
while (max_error > errorallow):
    iter = iter+1
    A = np.zeros((nQ,nQ))
    rhs = np.zeros(nQ)
    if(iter > maxiter): break

    for i in range(0,njunctions):
        for j in range(0,nQ):
            A[i,j] = junctionQ[i,j]
            rhs[i] += junctionQ[i,j]*Q[j]
        rhs[i] -= junctionO[i]

    for i in range(0,nloops):
        for j in range(0,nQ):
            A[i+njunctions,j] = loopQ[i,j]*k[j]*Q[j]*2
            rhs[i+njunctions] += loopQ[i,j]*k[j]*Q[j]**2
        rhs[i+njunctions] -= (loopH[i,0]-loopH[i,1])

    delQ = np.linalg.solve(A,rhs)
    Q -= delQ
    max_error = np.max(abs(delQ/Q))

print("iter = ",iter,". Discharges in the pipes are as
follows:")
print(Q)
```

Note, in particular, the data specified through the matrices `junctionQ`, `junc-tionO`, `loopQ`, and `loopH`. The first of these is a 4 × 7 matrix, with each row corresponding to a single junction of the system, and each element of a row representing the participation of the corresponding pipe. Hence, a value of zero (0) in a row is specified for those pipes that do not appear at the junction corresponding to the row. And, quite similar to the previous example, a value of +1 indicates an inflowing branch and a − 1 an outflowing branch, which is connected to the junction. The array `junctionO` specifies the outflows from the junctions and thus the number of elements of this array is equal to the number of junctions. The data specified in the array `junctionO` shown in the code above specify discharges of 0.10 and 0.05 m³/s for junctions 3 and 4. Those for junctions 1 and 2 are given as zero.

Similarly, the matrices `loopQ` and `loopH` are both specified as one row for each loop, whether open or closed. Specifically, the last row of these matrices traces the path of the open loop between the two given overhead tanks; while the two just above correspond to the two closed loops. Note that the values of the specified heads for the closed loops are given arbitrarily zero (0) values in the matrix `loopH`, since the energy balance equation starts and ends at the same location. However, specific values of 100m and 50m are specified for the two ends of the open loop of the system, the difference of which actually drives the flow through the system.

The data of the system of pipe networks (for the example of 7 pipes and 4 junctions) used as input to the Python program are similar to that of the previous example illustrating 3 pipes and 1 junction. The few differences that exist between the two are the following:

1. Since there are more than one junction in this example, the variable `junc-tionQ`, which was a one-dimensional array in the previous example, is now saved as a two-dimensional matrix, with the rows corresponding to the number of junctions (here, 4) and the columns corresponding to the total number of pipes in the system. The value of each element of the matrix is either 0, +1, or −1, which represents the absence (if the element is 0) or presence (+1 or −1) of the corresponding pipe at the junction, with +1 indicating a flow in the pipe toward the junction and −1 the opposite.

2. Outflows are specified at some of the junctions in this example and thus a one-dimensional array, `junctionO`, is used to input the specified outflow values. The array has the same number of elements as there are number of junctions and the elements of the array specify the value of the outflowing discharge at the corresponding junction (also called the demand at the junction). In the present example, out of four junctions, only two (junctions 3 and 4, shown in Figure 2.6) are imposed with outflow discharges. Thus, the first two elements of `junctionO` are zero, while the last two (corresponding to junctions 3 and 4) are indicated as 0.1 and 0.05, which are sample flow demands at these two junctions in m³/s.

For any other configuration of pipes, junctions, overhead tanks, or junction demand, the data in the program may be suitably modified.

The output of the code, when run, is as follows:

```
iter = 5. Discharges in the pipes (in m3/s) are as follows:
[0.50082705 0.35082705 0.20993843 0.29088862 0.04202442
0.20196285
 0.1488642]
```

## REFERENCES

Chapra, S. and Canale, R. (2021). *Numerical Methods for Engineers*. McGraw-Hill Education, 8th edition. https://www.mheducation.com/highered/product/numerical-methods-engineers-chapra-canale/M9781260232073.html

Chapra, S. C. (1997). *Surface Water Quality Modeling*. Waveland Press Inc. https://www.waveland.com/browse.php?t=378

Chaudhry, M. H. (2008). *Open Channel Flow*. Springer, 2nd edition.

Chow, V., Maidment, D. and Mays, L. (1988) *Applied Hydrology*. McGraw-Hill Book Company, New York.

Jain, M. K., Iyengar, S. R. K. and Jain, R. K. (2019).*Numerical Methods: for Scientific and Engineering Computation*. New Age International Private Limited, 7th edition. New Delhi.

Kopchenova, N. V. and Maron, I. A. (1981). *Computational Mathematics*. Mir Publishers, 1st English edition. Moscow.

Sherman, L.K. (1932). "Streamflow from Rainfall by Unit-Graph Method". *Eng. News Record*, 108, 501–505.

# 3 Ordinary Differential Equations

Many physical phenomena or processes occurring in the real world may be expressed mathematically as ordinary or partial differential equation(s), in contrast to others which may be defined by algebraic equations, as demonstrated in the Chapter 2. Some examples from the field of hydraulics, hydrology, contaminant transport, and water resources engineering that involve ordinary differential equations (ODEs) are presented in this chapter, which is divided into three sections. In the first section, different examples are discussed that fall into this category. The numerical techniques for solving the equations, and corresponding codes in Python appear in the following two sections, respectively. Systems involving partial differential equations (PDEs) are dealt with in the subsequent chapters.

## 3.1 EXAMPLES OF ORDINARY DIFFERENTIAL EQUATIONS IN HYDROLOGY, HYDRAULICS, AND WATER RESOURCES ENGINEERING

ODEs are encountered quite commonly in natural and engineered systems in the field of hydrology, hydraulics, and water resources engineering. Some examples have been discussed in the following section which provide a motivation for the rest of the content in this chapter, that is, techniques to solve ODEs and Python codes for implementing the techniques for computing the solutions.

### 3.1.1 EMPTYING OF A WATER TANK

We consider an initially empty water tank in the shape of an inverted truncated cone, much like a bucket, as shown in Figure 3.1. It is required to find the variation in the depth of water, $h$, of the tank if it is filled with a source of water, $Q_{in}$, and has a bottom outlet that also drains out a portion of the accumulating flow at a variable flow rate $Q_{out}$. Although $Q_{in}$ is possible to be controlled, the outflow from the tank $Q_{out}$ is solely dependent upon the water depth in the bucket, $h$.

We may assume the tank to be tapering uniformly, with the cross section at a height h, $A_h$, varying from that equal to the area at the bottom, $A_{bottom}$, up to $A_{top}$ at the top of the tank. $A_h$ may thus be expressed in terms of $A_{top}$ and $A_{bottom}$ by a linear function of the form:

$$A_h = A_{bottom} + \frac{A_{top} - A_{bottom}}{D} h \qquad (3.1)$$

DOI: 10.1201/9780429288579-3

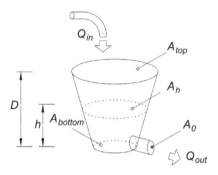

**FIGURE 3.1**    Water tank emptying and filling simultaneously.

If the cross section of the bottom outlet pipe is given as $A_o$ and the coefficient of discharge of the outlet pipe is known to be $C_d$, then the corresponding ODE may be set up as follows, using mass conservation laws:

$$A_t \frac{dh}{dt} = Q_{in} - Q_{out} \tag{3.2}$$

where $h$ is the depth of water at a given time in the tank and $Q_{in}$ may be constant or known as a function of time, that is, $Q_{in} \rightarrow Q_{in}(t)$. The outflow rate, $Q_{out}$, is dependent upon the pressure of water, the area of cross section of the outlet pipe, $A_o$, and the coefficient of discharge, $C_d$. That is, $Q_{out} = C_d\sqrt{2gh}A_o$. Equation (3.2) is an example of an ODE, and shows that it is possible to create our own differential equations if the physical background of the process that we wish to model is known.

### 3.1.2 COMPUTING FLOOD OUTFLOW FROM THE SPILLWAY OF A DAM BY THE LEVEL-POOL ROUTING METHOD

Reservoirs behind dams not only help in storing water for future use but also play a role in moderating floods passing through the dam and its spillway. The inflows reaching the reservoir of a typical dam (Figure 3.2) arrive from the contributing catchment and the excess that cannot be stored is released through the spillway (and sometimes through bottom outlets, which are not considered here).

The spillways, however, need to be designed for the worst possible flood and a "design flood hydrograph" is estimated from the hydrology of the catchment and rainfall inputs obtained from the hydro-meteorological studies of the region. Quite similar to the bucket flow problem discussed in the preceding section, the reservoir problem also requires the spillway outflow to be known for a given inflow hydrograph. By making use of the design flood hydrograph as the inflow to the reservoir, the outflow through the dam's spillway may be computed, together with the rise in the water level of the reservoir, by the following continuity equation:

$$\frac{dStorage}{dt} = Inflow - Outflow \tag{3.3}$$

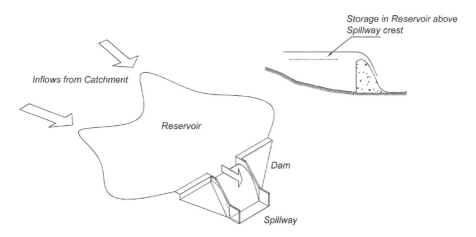

**FIGURE 3.2** An ungated spillway and reservoir of a typical gravity dam. Inset shows a section through the spillway.

The variable "storage", whose derivative appears on the left of Equation (3.3), is the volume of water that is stored in the reservoir above the crest level of the spillway (see inset in Figure 3.2). On the right hand side of the equation, the first term is the total flow into the reservoir from the catchment while the other is the discharge flowing out of the spillway. The equation is usually solved to calculate the peak outflow discharge, based upon which the spillway is designed. Also, the maximum rise in water level of the reservoir is important in understanding the possible extent of submergence that may occur in the river valley upstream of the dam. The computation process is generally known as the "level pool" method since the reservoir surface is considered to remain horizontal during the passage of the flood.

The inflow hydrograph is usually given in the form of a sequence of discharges at a given interval of time. It is usual, but not essential, for the outflows also to be computed at the same time step. In either case, the initial conditions are required to be specified. For example, the water in the reservoir may be assumed to be at the same level as of the crest of the spillway at the beginning of the computations.

Some other data that are required for completing the calculations include the storage volume in the reservoir above the spillway crest level and an assumed discharge capacity of the spillway. Usually, these data are provided in the form of a series of values, both corresponding to different elevations or heights measured above the spillway crest. Once the calculations are over, and the maximum rise in the water level of the reservoir is found out, it is checked whether the rise is within permissible limits. If not, then the spillway is increased in length and the calculations repeated to recheck the maximum rise of the reservoir-level, till a satisfactory length of the spillway is achieved.

### 3.1.3 WATER SURFACE PROFILE FOR STEADY-STATE GRADUALLY VARIED FLOWS

Water flowing through a channel or a river would have its free surface sloping up backwards if the channel slope is "mild", that is, when the normal depth of flow is greater than the critical depth for the given discharge. The flow would usually be subcritical and the free surface would conform to the typical M1 type of gradually varied flow profile, commonly observed behind reservoirs. It is often essential to compute the elevation profile of this "back-water" for checking the possible areas under submergence due to inundation caused by the construction of a dam.

As shown in the inset of Figure 3.3, if the water depth at a section of the channel is denoted as $y$, which is assumed to vary with the longitudinal distance $x$ measured upstream from the weir, the total energy head, $H$, may be written as (Chaudhry, 2008):

$$H = z + y + \alpha \frac{V^2}{2g} \tag{3.4}$$

In Equation (3.4), $H$ is the elevation of the energy grade line above an assumed datum; $z$ is the elevation of the channel bottom above the same datum; $V$ is the mean flow velocity, and $\alpha$ is the velocity-head coefficient, which we may approximate as 1.0 for simplicity. If the velocity, $V$, is expressed as $Q/A$ where $Q$ is the channel discharge and $A$ is the area of the cross section, then on differentiating Equation (3.4) with respect to the space variable $x$, the following governing differential equation is obtained:

$$\frac{dH}{dx} = \frac{dz}{dx} + \frac{dy}{dx} + \frac{Q^2}{2g} \frac{d}{dx}\left(\frac{1}{A^2}\right) \tag{3.5}$$

Equation (3.5) is an ODE of the first order and methods to solve this type of equation numerically have been discussed in a subsequent section. It may be noted that in order to completely solve a differential equation such as this the appropriate boundary conditions have to be defined. In this case, the steady-state discharge ($Q$) and the water level at the location of the weir are the necessary and sufficient conditions required to solve the equation completely.

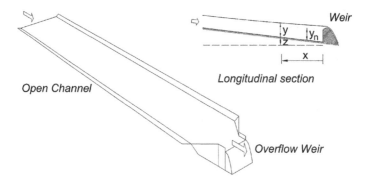

**FIGURE 3.3**   Gradually varied flow in an open channel. $y_n$ is the normal depth. Inset shows the backwater curve extending upstream.

### 3.1.4   STEADY-STATE CONCENTRATION PROFILE FOR DISSOLVED OXYGEN AND BIOCHEMICAL OXYGEN DEMAND IN ONE-DIMENSIONAL FLOWS

The water-quality parameters of dissolved oxygen ($DO$), and biochemical oxygen demand ($BOD$), are often considered as two of the primary indicators for understanding the health of a water body, such as a river or a lake. The first among these quantifies, as the name suggests, the amount of oxygen that is dissolved in the water. The oxygen in water, received from the atmosphere or aquatic plants, is important for the survival of the aquatic biota. $BOD$ is the amount of oxygen that is required by bacteria to react with and stabilize the organic waste load in water under aerobic conditions. Dispersion is commonly ignored while relating $DO$, $BOD$ with time and the reaction rates of the two parameters. The resulting equations, expressed as a pair of ODEs and originally suggested by Streeter and Phelps (1925), are usually stated in the forms given below:

$$\frac{d(BOD)}{dt} = -k_{deox}(BOD) \tag{3.6}$$

$$\frac{d(DOd)}{dt} = k_{deox}(BOD) - k_{aer}(DOd) \tag{3.7}$$

In Equations (3.6) and (3.7), the term $BOD$ is as stated above, while $DOd$ represents the dissolved oxygen deficit, which is the difference between the $DO$ value for complete saturation of oxygen and the $DO$ actually present in water. Both $BOD$ and $DOd$ (or for that matter, $DO$) are in same units, such as [ML$^{-3}$]. The reaction parameters, $k_{deox}$ and $k_{aer}$, represent the deoxygenation and reaeration rate constants, respectively, and both are in similar units of [T$^{-1}$].

Equations (3.6) and (3.7) provide the mathematical relation by which the $DO$ and $BOD$ at a location are calculated as a function of time, provided the initial values of the two are known at a point in time, and at a given location. The equations are both ODEs since time ($t$) is the only one independent parameter common to both equations. Quite often, the equations are converted in terms of the longitudinal distance variable (measured, say, along the length of a river), in order to predict the steady-state levels of $DO$ and $BOD$. By writing $d\cdots/dt$ as $\left(d\cdots/dx\right)\left(dx/dt\right)$, where $dx/dt$ is the velocity of the stream ($v$), Equations (3.6) and (3.7) may be rewritten as follows:

$$\frac{d(BOD)}{dx} = -\frac{k_{deox}}{v}(BOD) \tag{3.8}$$

$$\frac{d(DOd)}{dx} = \frac{k_{deox}}{v}(BOD) - \frac{k_{aer}}{v}(DOd) \tag{3.9}$$

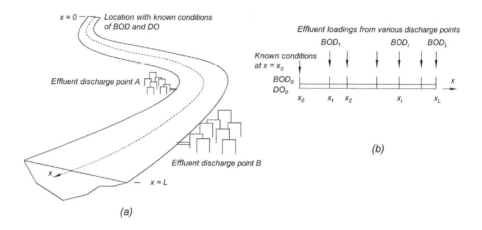

(b)

(a)

**FIGURE 3.4** (a) Schematic of a polluted river; (b) BOD "loadings" from effluent discharges.

Equations (3.8) and (3.9) too are ODEs, since each has only one independent variable – the distance variable $x$. Although these equations may be used to determine the distribution of *BOD* and *DO* along the length of the stream, they would be assumed to hold good for a steady-state condition in time. The pair of equations may be solved if the values of the two variables (*BOD* and *DO*) are known at an upstream point in the river. A schematic of a typical river with effluents discharged at specific locations is shown in Figure 3.4.

It is possible that along the stretch of a river, there are intermediate locations from where pollutants are released into the stream. These may also be incorporated into the system of equations (not directly, but in their integrated forms) if the *BOD* "loadings" are known equivalently from the discharging effluents.

## 3.1.5 Oscillations of Water Level in a Surge Tank

A surge tank of a hydropower system is a chamber located at the junction of the head-race tunnel, which is a mildly sloping conduit conveying water from the reservoir, and the penstocks, which are pressured conduits leading to the turbines, the flow being controlled through a valve. Schematically, a typical combination of the different components of the system is shown in Figure 3.5(a). At steady state, the inflowing water from the reservoir, as shown in the sectional view in Figure 3.5(b), equals the outflowing water to the turbine. Also, the water level in the surge tank remains at a steady or constant level at a depth, $h_f$, below the reservoir water level, where $h_f$ is equal to the frictional head loss in the tunnel. As often is the case, the head-race tunnel is long and the head loss due to friction is considered much larger than the entry and exit losses of the tunnel at its upstream and downstream ends. In that case, as shown in Figure 3.5(b), the two are ignored in comparison to $h_f$.

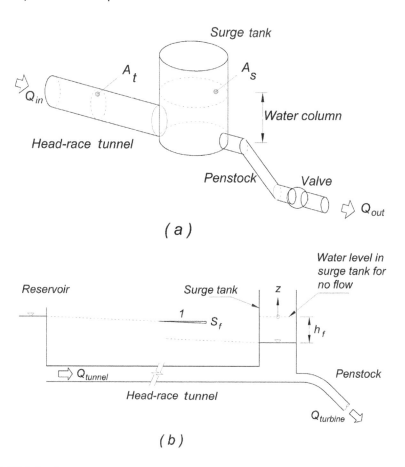

**FIGURE 3.5** Surge tank and other hydraulic components of a hydropower water conducting system. (a) Schematic view; (b) sectional view.

When the valve controlling the flow to the turbines is operated to reduce or increase the discharge to the turbines, the flow in the penstock changes, disturbing the steady-state balanced conditions discussed above. If the flow to the turbine ($Q_{turbine}$ in Figure 3.5b) is reduced, the flow coming in from the tunnel ($Q_{tunnel}$) is not able to immediately adjust to the lowered value because of the momentum of the comparatively large quantity of water in the tunnel. Thus, the rejected water from the turbine, that is, the difference in the two discharges, $Q_{tunnel}$ and $Q_{turbine}$, enters the surge tank, raising its level instantaneously. It takes some time to reach the new steady state, during which the water level in the surge tank oscillates sympathetically with the fluctuations in the flow of the tunnel. An opposite case may happen when the flow in the penstock is increased, in which case the flow in the tunnel is not able to adjust immediately because of the same reason as above and hence the water in the surge tank drops, supplying the immediate excess flow to the turbine.

Following Chaudhry (2008), we may describe the oscillations in the surge tank as a differential equation in terms of its water level, $z$, measured as positive upwards from the reference level as shown in Figure 3.5(b). This may be expressed as follows:

$$\frac{dz}{dt} = \frac{1}{A_s}\left(Q_{tunnel} - Q_{turbine}\right) \tag{3.10}$$

In the above, $A_s$ is the area of surge tank. At the same time, the flow in the tunnel, $Q_{turbine}$, also fluctuates and a second differential equation is introduced as given below:

$$\frac{dQ_{tunnel}}{dt} = \frac{gA_t}{L}\left(-z - \frac{fL}{2gD_tA_t^2}Q_{tunnel}^2\right) \tag{3.11}$$

In Equation (3.11), $L$, $D_t$, and $A_t$ are the length, diameter, and area of the tunnel, respectively. Also, $g$ is the acceleration due to gravity and $f$ is the friction factor for use in the Darcy–Weisbach equation for head loss. Thus, the variations in the water level of the surge tank or the discharge in the tunnel may be found out by solving the above pair of differential equations simultaneously for the given initial conditions.

### 3.1.6  RECHARGE OF RAINWATER INTO GROUND AND STEADY-STATE GROUNDWATER-TABLE PROFILE

In groundwater flow through an unconfined aquifer, a common problem is to determine the saturated seepage flow and the free surface profile of the seeping groundwater. The free surface is also known as the phreatic surface or groundwater table, and its knowledge is sometimes required for checking conditions of waterlogging. Groundwater flow takes place under the action of the potential difference existing at the ends of the domain. Further, the groundwater table could be recharged from infiltrating water from above, which may cause a seepage mound to form in the vicinity of the infiltration zone. Here too, a difference in heads is set up between adjacent groundwater zones causing a lateral movement of subsurface flow to occur.

A simplified example of groundwater seepage is shown in Figure 3.6, where a long strip of land is shown located between two water bodies having different water surface levels. The direction of seepage flow is from the higher to the lower head. There is also some amount of recharge occurring from above to add to the groundwater flow.

The assumptions essentially simplify the above situation to a one-dimensional problem. Assuming a unit width (1 m) thickness perpendicular to the paper, the following governing equation may be derived based upon the Dupuit assumptions (Wang and Anderson, 1982):

$$\frac{K}{2}\frac{d^2\left(h^2\right)}{dx^2} = -R \tag{3.12}$$

In Equation (3.12), $h$ is the saturated thickness of the aquifer [L], $x$ is the distance measured longitudinally [L], $R$ is the recharge rate [LT⁻¹], and $K$ is the hydraulic

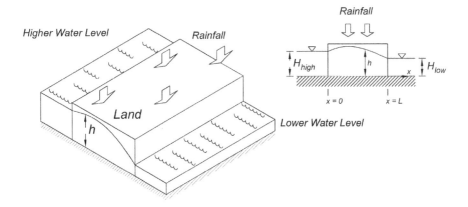

**FIGURE 3.6** Seepage flow and simultaneous recharge.

conductivity [$LT^{-1}$] of the aquifer material. Note that Equation (3.12) is non-linear because of the presence of $h^2$.

### 3.1.7 STEADY-STATE CONCENTRATION PROFILE FOR CONTAMINANT INJECTION IN ONE-DIMENSIONAL CHANNEL FLOWS

This problem relates to the transport of a non-conservative pollutant in a flowing stream or a river, which is continuously being injected at a given location within the water body. Here, we discuss only the steady-state distribution of the resulting pollutant concentration throughout the stream's length, which is commonly used for understanding the water quality variations in rivers and channels that are subjected to pollution load from adjoining sources (Chapra 1997; McCutcheon 1989; Thomann and Mueller 1987). The pollutant would be physically transported along the direction of flow in the stream by advection while it would attempt to spread in either direction due to dispersion. Note that the pollutant is non-conservative, that is, its concentration would change with time due to chemical reactions or biological evolutions.

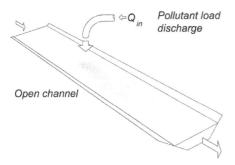

**FIGURE 3.7** Advection and dispersion of pollutant load in a steady open channel flow.

With the application of the principles of mass balance, advective transport, dispersive transport, and reaction kinetics, the following differential equation for steady state is obtained (Benedini and Tsakiris, 2013).

$$E\frac{d^2C}{dx^2} - v\frac{dC}{dx} - kC + S = 0 \tag{3.13}$$

Equation (3.13) contains the following variables: coefficient of dispersion $E$ [$L^2T^{-1}$]; steady-state velocity of flow $v$ [$LT^{-1}$]; reaction rate coefficient of decay $k$ [$T^{-1}$]; and the pollutant source $S$. Both $S$ and constituent concentration $C$ need to be in any compatible units. It is assumed that the flow velocity, $v$, is known. Or else, the steady-state flow equation for the channel has to be first solved and the velocity field ascertained. The dispersion coefficient, $E$, and reaction coefficient $K$ are also generally understood to have been estimated from field observations or from standard tables. The only unknown variable in Equation (3.13), therefore, is the constituent concentration $C$, which varies as a function of the longitudinal distance along the channel in both directions. As boundary conditions, the concentration of the constituent, $C$, or its spatial gradient, $dC/dx$, has to be specified. Specifying the value of $C$, amounts to implementing a Dirichlet condition, while specifying its gradient or derivative is the application of a Neumann condition. Since Equation (3.13) is a second-order ODE, two such conditions may be specified, and it could be a combination of both Dirichlet and Neumann conditions. Note that any additional source carrying specified concentrations released into the flow have to be defined along with their locations for correctly obtaining the solution to such a problem.

## 3.2  SOLUTION TECHNIQUES

In an ODE, there exists only one independent variable and one dependent variable. However, depending upon the highest derivative appearing in the equation, the ODEs may be classified as either first, second, or higher order. In this chapter, the examples of ODE-s considered are of the first and second order. Those of first-order type are the following:

1. Simultaneous filling and emptying of a water from a tank
2. Routing of a flood hydrograph through the spillway of a dam
3. Profile of water surface in steady-state gradually varied flows
4. Concentration profiles of *BOD* and *DO* along the length of a stream

While the following are examples of second-order differential equations:

5. Oscillations of the water level in a surge tank
6. Groundwater-table profile from recharge due to rainfall
7. Steady-state concentration profile for point loading in one-dimensional flows

Of the above cases, the time-dependent phenomena (examples 1, 2, and 5) are also called "initial value" problems, since the values of the dependent variable(s) at the beginning or an initial state needs to be specified. Examples 1 and 2, being of the first-order type of ODE require only one starting condition to be specified initially while example 5, being of second-order, requires two boundary conditions to be specified at the initial time for obtaining solutions at future times. Examples 3 and 4 are space-dependent single-point "boundary-value" problems, in which a known condition is specified at one of the ends of the one-dimensional computation domain. Examples 6 and 7 are two-point "boundary-value" problems, where the known values are provided at either end of the domain. The numerical schemes for solving the equations are discussed in the following sections.

## 3.2.1   FIRST-ORDER ORDINARY DIFFERENTIAL EQUATIONS

The general first-order ordinary differential equations subject to an initial condition may be written as:

$$\frac{dy}{dx} = f(x, y) \text{ with } y(x_0) = y_0 \tag{3.14}$$

where $f(x, y)$ is any function of $x$ and $y$. The equation may be non-linear depending upon the nature of the function $f(x, y)$. The numerical methods for solving equations of the form given in the above differential equation start with an initial condition $(x_0, y_0)$ and progress in small steps along the x-axis. The successive values of the step points are designated as $x_1$, $x_2$, etc., and the $i^{th}$ step may be designated as $x_i$. The step size or step length $h (= x_{i+1} - x_i)$ is specified by the user and at each $x_i$, the corresponding value of abscissa $y_i$ is evaluated by following an algorithm such as:

$$y_{i+1} = y_i + \int_{x_i}^{x_{i+1}} \frac{dy}{dx} dx \tag{3.15}$$

Of the many approaches to solve Equation (3.15), we shall discuss the very basic Euler's method or its modified form, the Heun's method, and the more accurate and widely used fourth-order Runge–Kutta method.

### 3.2.1.1   Euler's Method

This technique is one of the simplest for solving ordinary initial-value differential equations. Although the basic Euler's method is not accurate enough, it forms the basis of other methods in the same category. The numerical integration of Equation (3.15) is performed by evaluating the derivative at the initial condition. By knowing the value $y_0$ at $x_0$, as given in the condition of Equation (3.14), the Euler's method estimates the value of $y$ at the location $(x_0 + h)$ as follows:

$$y_1 = y_0 + h f(x_0, y_0) \tag{3.16}$$

With the value of the function $y_1$ found, the procedure is repeated for subsequent values in a similar sequence as:

$$y_{i+1} = y_i + h f\left(x_i, y_i\right) \tag{3.17}$$

The algorithm for implementing the above steps may be written as:

```
define f(x,y)
definexinitial, xfinal (starting and ending values of
  independent variables)
define h (step size)
xi = xstart
while xi+1< xfinal
      y0 = h·f(xi,yi)
      Yi+1 = yi+y0
      xi+1 = xi+h
end while
```

The above basic Euler's method may be improved in accuracy by evaluating the function $f \left(= \dfrac{dy}{dx}\right)$ at both the points $x_{i+1}$ and $x_i$ and taking an average of the two. Also known as the modified Euler's method or the Heun's method, the improved value of the function by this procedure is found out as:

$$y_{i+1} = y_i + h \frac{f\left(x_i, y_i\right) + f\left(x_{i+1}, y_{i+1}^*\right)}{2} \tag{3.18}$$

where $y_{i+1}^*$, the first stage approximation of the function evaluated using the simple Euler's method, is given by

$$y_{i+1}^* = y_i + h f\left(x_i, y_i\right) \tag{3.19}$$

The algorithm for this step may be written as below:

```
define f(x,y)
define xinitial, xfinal define h (step size)
xi = xstart
while xi+1 < xfinal
      xi+1 = xi+h
      y* = yi+h·f(xi+1,yi)
      f* = f(xi+1,yi+y*)
      Yi+1 = yi+ h·(f+f*)/2
end while
```

### 3.2.1.2 Fourth-Order Runge–Kutta Method

Euler's and Heun's methods are both versions of a broader class of technique that goes by the name of Runge–Kutta (Chapra and Canale 2021) for solving ordinary initial-value differential equations. Many versions of the Runge–Kutta (RK) method exist, but all can be expressed in the generalized form:

$$y_{i+1} = y_i + \varphi(x_i, y_i, h) \cdot h \tag{3.20}$$

where $\varphi(x_i, y_i, h)$ is called an increment function and can be interpreted as an average slope over the step interval. The fourth-order RK method (sometimes written as the RK4 method) is especially popular and is expressed as:

$$K_0 = h \cdot f(x_i, y_i) \tag{3.21a}$$

$$K_1 = h \cdot f\left(x_i + \frac{1}{2}h, \ y_i + \frac{1}{2}K_0\right) \tag{3.21b}$$

$$K_2 = h \cdot f\left(x_i + \frac{1}{2}h, \ y_i + \frac{1}{2}K_1\right) \tag{3.21c}$$

$$K_2 = h \cdot f\left(x_i + \frac{1}{2}h, \ y_i + \frac{1}{2}K_2\right) \tag{3.21d}$$

$$y_{i+1} = y_0 + \frac{1}{6} \cdot (K_0 + 2K_1 + 2K_2 + K_3) \tag{3.21e}$$

To initiate an iterative solution to the above set of equations, the function $f(x_i, y_i)$ and the initial condition $y(x_0) = y_0$, provided by Equation (3.14), are used.

The algorithm for the fourth-order RK method may be expressed in the following form:

```
define f(x,y)
define x_initial, x_final define h (step size)
x_i = x_start
while x_{i+1} < x_final
        y0 = h·f(x_i,y_i)
        y1 = h·f(x_i+0.5h,y_i+0.5y0)
        y2 = h·f(x_i+h,y_i+0.5y1)
        y3 = h·f(x_i+0.5h,y_i+0.5y2)
        Y_{i+1} = y_i+(y0+2·y1+2·y2+y3)/6
        x_{i+1} = x_i+h
end while
```

### 3.2.1.3 Accuracy and Stability

It may be noted that Euler's method is obtained from the Taylor' series, as given below:

$$y(x_0 + h) = y(x_0) + h\frac{dy(x_0)}{dx} + \frac{h^2}{2!}\frac{d^2y(x_0)}{dx^2} + \frac{h^3}{3!}\frac{d^3y(x_0)}{dx^3} + \ldots \tag{3.22}$$

The terms following the first two terms on the right hand side of the above equation are called the error terms and are truncated to obtain Equation (3.16), in which $h$ is the step size. Note that the function $f$ in Equation (3.16) is the first derivative in Equation (3.22). In assessing the accuracy of the method, the dominant error term gives useful information regarding the influence of the step size $h$. The dominant term is the term with the lowest power of $h$ in the error terms of the Taylor series. From Equation (3.22), the dominant error-term for Euler's method is seen to be $\dfrac{h^2}{2!}\dfrac{d^2 y(x_0)}{dx^2}$. The $h^2$ term implies that if the step length is reduced by a factor of 2, the local error (that is, the error for one step) will be reduced by a factor of 4. The overall error accumulated over the total number of steps, also known as global error, would be $h^2$ multiplied by $1/h$ since the number of steps is proportional to $1/h$. Thus, the global error for Euler's method is of the order of $h$.

For analysing the order of error in the Heun's method, we note that in the Taylor series expansion (Equation 3.22), we are indirectly truncating the terms beyond the third term as follows:

$$y(x_{i+1}) = y(x_i) + hy'_{xi} + \frac{h^2}{2!}\left\{ \frac{y'_{xi+1} - y'_{xi}}{h} + O(h) \right\} + O(h^3) \qquad (3.23)$$

In Equation (3.23), the notation $y'$ is used for $\dfrac{dy}{dx}$ and $O(h)$ is the order of error on neglecting higher-order terms for the second derivative in the third term and $O(h^3)$ is the order of error on neglecting the third-order derivatives and remaining terms in the Taylor series. Equation (3.23) thus becomes:

$$y(x_{i+1}) = y(x_i) + \frac{h}{2}\left[ y'_{xi} + y'_{xi+1} \right] + O(h^3) \qquad (3.24)$$

Thus, the local error for the Heun's method is of the order of $h^3$. Here too, the global error would reduce by an order to $h^2$ since the local error would be accumulated for each step (incremented by $h$) and the number of steps is proportional to $1/h$. Similar analyses show that the global error for the fourth-order RK method is of the order of $h^4$. Because of its relatively high accuracy, this method is widely used for solving ordinary initial-value PDEs.

## 3.2.2   SECOND-ORDER ORDINARY DIFFERENTIAL EQUATIONS

In order to solve problems of such types, the single higher-order equation may be reframed as a system of more than one ODEs, which need to be solved simultaneously. For example, consider a generic second-order differential equation as given below:

$$\frac{d^2 y}{dx^2} = f(x,y) \text{ with } y(x_0) = y_0 \text{ and } \left.\frac{dy}{dx}\right|_{x_0} = y'_0 \qquad (3.25)$$

where $f(x, y)$ is any function of $x$ and $y$. Also, the given second-order equation could be non-linear. The initial conditions specified are two now, since the differential equation is of the second order. For progressing in steps along the $x$-axis, the $i^{th}$ step may be designated as $x_i$, with the step size $h$ ($=x_{i+1} - x_i$) defined by the user.

The second-order equation is reduced to a pair of first-order equations by introducing a new variable, say $z$, as under

$$z = \frac{dy}{dx} \qquad (3.26)$$

and a derivative for the variable $z$, such as:

$$z' = \frac{d^2 y}{dx^2} = f(x, y) \qquad (3.27)$$

From here on, the Euler's, Heun's, RK4 or any other suitable numerical method may be employed by solving Equations (3.26) and (3.27) simultaneously over each step of computation, that is, moving from the known variables at $x = x_i$ to finding the unknowns at $x = x_{i+1}$.

### 3.2.3 Two-Point Boundary Value Problems

The examples considered in this section on two-point boundary value problems are both second-order, defined in a one-dimensional spatial domain. In general, the equations may be expressed in the following form:

$$\frac{d^2 y}{dx^2} + p(x)\frac{dy}{dx} + q(x) + r = 0 \qquad (3.28)$$

We adopt the second-order central finite-difference approximations for solving Equation (3.28) in the following way:

$$\frac{y_{i+1} - 2y_i + y_{i-1}}{h^2} + p_i \frac{y_{i+1} - y_{i-1}}{2h} + q_i y_i + r_i + O(h^2) = 0 \qquad (3.29)$$

For the purpose of computations, the truncation error is omitted. For $N$ computation points, $N-2$ equations similar to above are generated. Hence, in order to solve for the unknowns completely, two more equations are additionally required. These may be obtained from the two boundary conditions, which may be of the Dirichlet or Neumann types. In case of the former, the values given are directly imposed in the form of two equations. For the latter, with a specified gradient equal to zero, two consecutive grid point variables at the boundary are equated.

## 3.3 PYTHON PROGRAMS

Programs in Python for solving the examples discussed in the previous sections on surface and groundwater hydrology, hydraulics, and contaminant transport in one-dimensional channels, expressed using ODEs and solved by different numerical techniques, are presented below.

### 3.3.1 First-Order ODE: Solving the Tank Filling and Emptying Problem Using Heun's Method

Section 3.1.1 discussed the problem of the storage of water in a tank, shaped in the form of a truncated cone and having uniformly increasing cross section from bottom to top (Figure 3.1). Water enters from above through a tap but is simultaneously drained out through an outlet at the bottom of the tank. The governing differential equation for the problem, Equation (3.2), is reproduced below:

$$A_t \frac{dh}{dt} = Q_{in} - Q_{out} \tag{3.2}$$

In Equation (3.2), $h$ denotes the depth of water in the tank at a given time which is sought to be computed at subsequent time steps assuming an initial value, say $h_o$, defined at the beginning, say at $t = 0$. Considering the level of the outlet as the datum, we may assume the initial depth of water ($h_o$) with respect to this datum to be zero if the tank is initially empty. The independent variable here is time, $t$, which may be counted from zero and incremented at intervals of a specified time step, $dt$. In Equation (3.2), $A_t$ is the area of the free surface of water in the tank at a given time, $t$. Values of the inflow discharge, $Q_{in}$, is assumed to be known at any given time. The depth of the tank may be taken as $D$ with the cross-sectional area assumed to increase uniformly from bottom upward. It is further assumed that the cross sections are specified at the bottom and the top of the tank, from which the variation of the area with depth may be established. The area of the outlet pipe, $A_o$, and its coefficient of discharge, $C_d$ are also assumed to be known. The outlet discharge may then be expressed as $Q_{out} = C_d \sqrt{2gh} A_o$.

The Python code for the problem, implementing the Heun's method of integration, is given below.

```
# Flow through a truncated conical water tank by Heun's
# method
import numpy as np
import matplotlib.pylab as plt

g = 9.81
Atop = 1.0
Abottom = 0.5
D = 0.75
Aoutlet = 0.005
Cd = 0.7
time_simulation = 600
time_control1, time_control2 = 300,510
Qin1, Qin2 = 0.01,0.005
```

```
dt = 1     # Time step (in seconds)
ntimes = int(time_simulation/dt)
print(ntimes)
h=np.zeros(ntimes)   # Sequence of water depths in bucket
time = np.zeros(ntimes)   # Time markers for plotting

h[0] = 0.0
for n in range (0,ntimes-1):
    time[n+1] = time[n] + dt
    if(time[n] <= time_control1):
        Qin = Qin1
    elif(time[n] > time_control1 and time[n] <=
      time_control2):
        Qin = Qin2
    else:
        Qin = 0.0
    Qout = Cd*Aoutlet*np.sqrt(2*g*h[n])
    Area = Abottom + h[n]*(Atop-Abottom)/D
    dhdt0 = (Qin - Qout)/Area
    h1 = h[n]+dhdt0*dt
    Qout1 = Cd*Aoutlet*np.sqrt(2*g*h1)
    Area1 = Abottom + h1*(Atop-Abottom)/D
    dhdt1 = (Qin - Qout1)/Area1
    dhdt = 0.5*(dhdt0+dhdt1)
    h[n+1] = dhdt*dt+h[n]
    if(h[n+1] < 0.01): h[n+1] = 0.0

fig = plt.figure()
ax = fig.add_subplot(1, 1, 1)
ax.plot(time, h)
ax.set_xlabel('Time (s)')
ax.set_ylabel('Water Depth (m)')
plt.show()
```

The variables used in this program are described below:

| Variable | Description | Variable | Description |
|---|---|---|---|
| g | Acceleration due to gravity $(m/s^2)$ | Aoutlet | Cross-sectional area of outlet vent or pipe $(m^2)$ |
| Atop | Cross-sectional area at the top of tank $(m^2)$ | Cd | Coefficient of discharge (-) |
| Abottom | Cross-sectional area at the bottom of tank $(m^2)$ | time_ simulation | Total simulation time (s) |
| D | Depth of tank, between Atop and Abottom (m) | dt | Time step used for computations (s) |
| time_ control1, time_ control2 | Two time (s) points for specifying inflow discharges (Qin1, Qin2) | Qin1, Qin2 | Two values of inflow discharges $(m^3/s)$, changing with time |

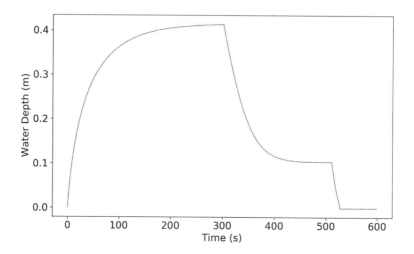

**FIGURE 3.8** Variation of the depth of water in a tank with time. Inflow data are given as 0.01 m³/s from beginning to 300 s, and from then on at 0.005 m³/s till 510 s, and zero beyond 510 s. (Color image available in eBook).

On running the above program in Python with the data specified in lines 6 to 15, the graphical output of Figure 3.8 is obtained.

### 3.3.2 FIRST-ORDER ODE: FLOOD ROUTING THROUGH A RESERVOIR AND SPILLWAY USING HEUN'S METHOD

The governing equation for routing a flood hydrograph through the spillway of a dam is given by Equation (3.3), which is repeated below for convenience:

$$\frac{dStorage}{dt} = Inflow - Outflow \qquad (3.3)$$

Notice that the Inflow hydrograph is given at discrete times, usually at a constant interval of time step $\Delta t$. In Equation (3.3), both *Storage* and *Outflows* are unknown, but they are related to one another through the common variable $h$. The equation is rewritten as follows:

$$\frac{dStorage}{dt} = \frac{dStorage}{dh}\frac{dh}{dt} = Inflow - Outflow \qquad (3.30)$$

From which, we may write an equivalent form as:

$$\frac{dh}{dt} = \frac{\left(Inflow - Outflow\right)}{\left(\dfrac{dStorage}{dh}\right)} \qquad (3.31)$$

   In Equation (3.31), only the time rate of change of the depth $h$, appears on the left of the equation, which is solved for $h$ using the Heun's method. The more accurate fourth-order Runge–Kutta method may be used instead. We need to keep the computation time step smaller than the time step at which the inflows have been specified, or at most, equal to it. Knowing $h$, the outflow may be computed at any time step along with the storage from the given relations between $h$ and the two variables. The following Python program incorporates the necessary lines of code for solving the above problem.

```python
# Reservoir routing by the level pool method
import numpy as np
import matplotlib.pylab as plt
from numpy import array

H = array([0.0,1.0,2.0,3.0,4.0,5.0,6.0,7.0,8.0,9.0])
S = array([0.0,6.25,9.50,11.25,12.75,14.0,15.0,15.75,
   16.25,17.0])
Q = array([0,100,300,500,800,1200,1620,2040,2450,2970])
I = array([0,100,250,500,1625,1750,1500,1000,500,350,
   250,175,100,50,25,0,0,0,0,0])
Dt = 1.0*3600 # Time step for inflow input (in seconds)
dt = 0.2*3600 # Time step for computations (in seconds)
errorallow = 0.001
maxiter = 10

ndepths = len(H)
ntimes = len(I)
time_simulation = (ntimes-1)*Dt # (in seconds)
nt = int(time_simulation/dt)
h = np.zeros(nt)
In = np.zeros(nt)
Out = np.zeros(nt)
time = np.zeros(nt)

def dSdh(depth):
    for i in range(0,ndepths):
        if(depth >= H[i] and depth < H[i+1]):
            dSdh = (S[i+1]-S[i])/(H[i+1]-H[i])*1000000
            return dSdh

def Outflow(depth):
    for i in range(0,ndepths):
        if(depth >= H[i] and depth < H[i+1]):
            Outflow = Q[i]+(Q[i+1]-Q[i])/(H[i+1]-H[i])*
   (depth-H[i])
```

```
            return Outflow

def Inflow(t):
    for i in range(0,ntimes):
        t1 = i*Dt
        t2 = (i+1)*Dt
        if(t >= t1 and t < t2):
            Inflow = I[i]+(I[i+1]-I[i])/Dt*(t-t1)
            return Inflow

h2 = 0.0
for n in range (1,nt):
    time[n] = time[n-1]+dt
    I1 = In[n] = Inflow(time[n])
    h1 = h2
    O1 = Outflow(h1)
    S1 = dSdh(h1)
    iter = 0
    maxerror = 1.0
    dhdt1 = (I1-O1)/S1
    h2star = h1+dhdt1*dt
    while(maxerror > errorallow):
        iter = iter+1
        if(iter > maxiter): break
        O2star = Outflow(h2star)
        S2star = dSdh(h2star)
        dhdt2 = (I1-O2star)/S2star
        h2 = h1+(dhdt1+dhdt2)/2*dt
        O2 = Outflow(h2)
        maxerror = abs(h2-h2star)
        h2star = h2
    Out[n] = O2
    h[n] = h2

fig = plt.figure()
ax = fig.add_subplot(1, 1, 1)
ax.plot(time/3600, In,'bo-')
ax.plot(time/3600, Out,'yo-')
ax.set_xlabel('Time (hours)')
ax.set_ylabel('Inflow / Outflow (m3/s)')
plt.show()
```

The variables used in this program are described below:

| Variable | Description | Variable | Description |
| --- | --- | --- | --- |
| H | Array of discrete values of depth, above spillway crest (m) | Dt | Time step for specifying inflow discharges, I (s) |
| S | Array of storage values corresponding to the depths in H (Mm³) | dt | Time step used for computations (s) |
| Q | Array of outflows corresponding to the depths in H (m³/s) | errorallow | Error tolerance value for iterative computation of h (m) |
| I | Array of inflow discharges (m³/s) | time_simulation | Total simulation time (s) |
| ntimes | The number of time points with specified inflows | nt | Number of time steps for computation |

Note that as the computations progress in time, the solutions are obtained through iterations within each time step. This is because both the storage and the spillway discharge capacity data are assumed as linearly piecewise continuous functions, specified at discrete elevations, and not as closed-form functions. Thus, there are possibilities of non-convergence of the desired result, if a direct solution is attempted at each time step. Despite this, non-convergent oscillations may still occur for the solution variables and such conditions may be avoided by breaking out of the iterative loop beyond a given number of iterations. On running the above program with the data specified in lines 6 to 15, the following graphical output (Figure 3.9) is obtained.

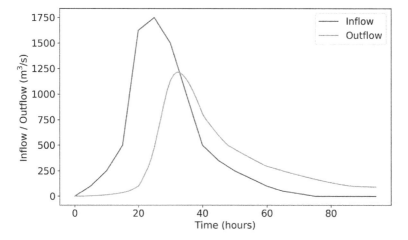

**FIGURE 3.9** Inflow hydrograph (in blue) and outflow (in orange) plotted against time. (Color image available in eBook).

### 3.3.3  First-Order ODE: Computation of the Back-Water Gradually Varied Flow Profile Using Fourth-Order Runge–Kutta (RK4) Method

The governing differential equation for the steady-state gradually varied flow profile behind a dam, also known as the backwater (Figure 3.3), is given by Equation (3.5), which is reproduced here:

$$\frac{dH}{dx} = \frac{dz}{dx} + \frac{dy}{dx} + \frac{Q^2}{2g}\frac{d}{dx}\left(\frac{1}{A^2}\right) \tag{3.5}$$

In the above equation, $H$, the elevation of the energy grade line measured above a datum level, is the only unknown variable that varies with the distance $x$, measured upstream from the dam. The channel is considered trapezoidal, with specified bed width and side slope. Also, the depth of water in the reservoir just behind the weir or the spillway of the dam is assumed to be known along with the discharge coming into the reservoir. The friction roughness factor of the channel is specified in terms of the Manning's roughness coefficient. The problem is solved using the RK4 method, treating it as a boundary value problem. The corresponding Python program is described below.

```
# Backwater curve profile in trapezoidal channel using
   the 4th-Order R-K method
import numpy as np
import matplotlib.pylab as plt

g=9.81
b0=40.0
s=2.0
s0=0.001
mn=0.035
Q=4000
yd=25.0
length=20000
npoints=101
x = -1.0*np.linspace(0, length, npoints)
dx=x[1]-x[0]

y=np.ones(npoints)*yd
z=-1.0*s0*x

def area(y):
    area=y*(b0+s*y)
    return area

def wetperi(y):
    wetperi=b0+2*y*np.sqrt(1+s*s)
```

```
    return wetperi

def topwidth(y):
    topwidth=b0+2*s*y
    return topwidth

def slopefriction(manning,qq,aa,rr):
    slopefriction=(manning*qq)**2/(aa*aa*rr**1.333)
    return slopefriction

v=np.ones(npoints)*Q/area(yd)

for i in range(0,npoints-1):
    y1=y[i]
    a1=area(y1)
    p1=wetperi(y1)
    r1=a1/p1
    b1=topwidth(y1)
    sf1=slopefriction(mn,Q,a1,r1)
    k1=(s0-sf1)/(1-(b1*Q*Q)/(g*a1**3))
    x2=x[i]+0.5*dx
    y2=y[i]+0.5*k1*dx
    a2=area(y2)
    p2=wetperi(y2)
    r2=a2/p2
    b2=topwidth(y2)
    sf2=slopefriction(mn,Q,a2,r2)
    k2=(s0-sf2)/(1-(b2*Q*Q)/(g*a2**3))
    x3=x[i]+0.5*dx
    y3=y[i]+0.5*k2*dx
    a3=area(y3)
    p3=wetperi(y3)
    r3=a3/p3
    b3=topwidth(y3)
    sf3=slopefriction(mn,Q,a3,r3)
    k3=(s0-sf3)/(1-(b3*Q*Q)/(g*a3**3))
    x4=x[i]+dx
    y4=y[i]+k3*dx
    a4=area(y4)
    p4=wetperi(y4)
    r4=a4/p4
    b4=topwidth(y4)
    sf4=slopefriction(mn,Q,a4,r4)
    k4=(s0-sf4)/(1-(b4*Q*Q)/(g*a4**3))
    yfinal=y[i]+1/6*(k1+2*k2+2*k3+k4)*dx
    y[i+1]=yfinal
    v[i+1]=Q/area(y[i+1])
```

```
H=y+z
fig = plt.figure()
ax = fig.add_subplot(1, 1, 1)
ax.plot(x/1000, H)
ax.plot(x/1000, z)
ax.set_xlabel('Distance from dam (km)')
ax.set_ylabel('Water surface and channel bed (m)')
plt.show()
np.savetxt('GVF.v.csv', v , delimiter=',')
```

The variables used in this program are described below:

| Variable | Description | Variable | Description |
|---|---|---|---|
| g | Acceleration due to gravity (m/s$^2$) | mn | Manning's friction coefficient (m$^{1/3}$/s) |
| b0 | Bottom width of the trapezoidal channel (m) | Q | Inflowing discharge (m$^3$/s) |
| s | Side slope of the banks of the channel (1V:'s'H) | yd | Depth of reservoir upstream of dam (m) |
| s0 | Bed slope = tan(theta); theta = angle of bed with horizontal | length | Length of channel for which backwater profile is calculated (m) |
| GVF.v.csv | File for saving results. To be used for other programs. | npoints | Number of space points for dividing the channel uniformly and used to find computation step |

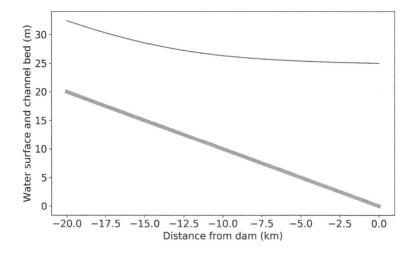

**FIGURE 3.10**  Gradually varied flow profile (blue) and channel bed (orange) for the backwater behind a dam. Flow is from left to right. The weir or a spillway of the dam exists at the right boundary. (Color image available in eBook).

On running the above program in Python, with the data specified in lines 6 to 14, the graphical output of Figure 3.10 is obtained. Note that the computed velocities are stored in the file GVF.v.csv for later use in another program, discussed below.

### 3.3.4 First-Order ODE: Computing the Steady-State BOD and DO Concentration Profiles in a One-Dimensional Stream Using Heun's Method

The *BOD* and *DO* in a stream may be computed using Equations (3.8) and (3.9) reproduced once more below:

$$\frac{d(BOD)}{dx} = -\frac{k_{deox}}{v}(BOD) \tag{3.8}$$

$$\frac{d(DOd)}{dx} = \frac{k_{deox}}{v}(BOD) - \frac{k_{aer}}{v}(DOd) \tag{3.9}$$

In the pair of equations, the velocity of the stream ($v$) is assumed to be known at all locations. Also, the two variables appearing in the equations are the BOD and the dissolved oxygen deficit (written as *DOd* in the equations), which is the difference between the maximum possible saturated oxygen concentration (*Dsat*) of the water at the given temperature and pressure and that which is actually existing in the stream. Thus, after solving Equations (3.8) and (3.9) simultaneously to find *BOD* and *DOd*, the actual dissolved oxygen, *DO*, is found by equating it to *Dsat – Dod*.

Bungay (1998) presents a code in the BASIC programming language for solving the coupled differential equations. Here, a similar algorithm is used in the following Python program which employs the Heun's method to solve the above system of equations, assuming that the upstream boundary values of BOD and DO are specified in the problem. Further, an array is created for specifying any intermediate pollutant loads. The data used in the Python code below are not taken directly from any field-problem, but based on the pollution in the river Yamuna flowing past the national capital region of Delhi, India, which is severe up to Kanpur (a distance of about 500 km, with an average river velocity of 0.5 ms[-1]). Some references provide the magnitude of different parameters which conform to the range of the data assumed here (for example, CPCB 2006; Jaiswal et al. 2019; Sharma and Singh 2009).

```
# DO - BOD curves solved by Heun's method
import numpy as np
import matplotlib.pylab as plt

length = 500000 # meters
npoints = 101
Kdox = 0.40 # per day
```

```
Kaer = 0.50 # per day
Dsat = 8 # mg/m3
B0 = 4.0 # mg/m3
D0 = 1.0 # mg/m3
v = 0.50 # m/s

x = np.linspace(0, length, npoints)
dx = x[1]-x[0]

B = np.zeros(npoints)
B[20] = 10.0; B[70] = 5.0
BOD = np.zeros(npoints)
DOd = np.zeros(npoints)
DO = np.zeros(npoints)
B1 = BOD[0] = B0
D1 = DOd[0] = D0
DO[0] = Dsat-DOd[0]

for n in range (1,npoints):
    Kd = (Kdox/v)/86400
    Ka = (Kaer/v)/86400
    dBdx1 = (-1.0)*Kd*B1
    dDdx1 = Kd*B1-Ka*D1
    B2star = B1+dBdx1*dx
    D2star = D1+dDdx1*dx
    dBdx2 = (-1.0)*Kd*B2star
    dDdx2 = Kd*B2star-Ka*D2star
    B2 = B1+dx*(dBdx1+dBdx2)/2
    D2 = D1+dx*(dDdx1+dDdx2)/2
    BOD[n] = B1 = B2+B[n]
    DOd[n] = D1 = D2
    DO[n] = Dsat-DOd[n]

fig = plt.figure()
ax = fig.add_subplot(1, 1, 1)
ax.plot(x/1000, BOD, label='BOD')
ax.plot(x/1000, DOd, label='DO deficit')
ax.plot(x/1000, DO, linewidth=4, label='DO')
ax.set_xlabel('Distance (km)')
ax.set_ylabel('Concentration (g/m3)')
leg = ax.legend()
plt.show()
```

The variables used in this program are described below:

| Variable | Description | Variable | Description |
|---|---|---|---|
| length | Total length of the stream considered (m) | npoints | Number of space points for dividing the channel uniformly and used for computing depth y |
| Kdox | Deoxygenation rate constant (per day) | B0 | BOD value and the upper end of the stream |
| Kaer | Reaeration rate constant (per day) | D0 | DO value and the upper end of the stream |
| BOD | Array for computing BOD for the npoints | Dsat | Saturated dissolved oxygen concentration [ML$^{-3}$] like (g/m$^3$) |
| DOd | Array for computing DO deficit for the npoints | B | Array for specifying point loading. Default zero values to be changed to provide BOD data. |

On running the above program, the graphical output (Figure 3.11) shows the steady-state distribution of *BOD*, *DO* deficit, and *DO* along the stretch of the river considered. Point loading of pollutants may be specified through the B array (which is initialized to zero at the beginning). In the above program, two such point values of BOD are specified as B[20] and B[70]. Since the length of the computational domain is 500 km, these two loading points lie at 100 km and 350 km, respectively, downstream of the origin.

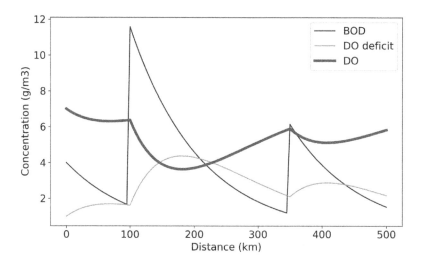

**FIGURE 3.11** Channel length-wise profile of BOD (blue), DO deficit (orange), and DO (green-bold) for 500 km hypothetical channel with contaminant loadings at two intermediate points. (Color image available in eBook).

### 3.3.5  SECOND-ORDER ODE: SURGE-TANK OSCILLATION PROBLEM SOLVED USING HEUN'S METHOD

The pair of governing equations for oscillations of the water body in a surge tank, discussed in Section 3.1.5, are reproduced below:

$$\frac{dz}{dt} = \frac{1}{A_s}\left(Q_{tunnel} - Q_{turbine}\right) \tag{3.10}$$

$$\frac{dQ_{tunnel}}{dt} = \frac{gA_t}{L}\left(-z - \frac{fL}{2gD_tA_t^2}Q_{tunnel}^2\right) \tag{3.11}$$

The depth of water in the tank at any instant of time, $t$, is $h$, which may be initially considered at a steady-state level. Other parameters in the equation, $A_t$, $A_s$, $L$, $D$, $D_t$, $f$, and $g$ are, respectively, cross-sectional area of the head-race tunnel, that of the surge tank, length of the tunnel, its diameter, flow friction factor for the tunnel, and acceleration due to gravity.

```
# Surge tank oscillations solved by Heun's method
import numpy as np
import matplotlib.pylab as plt

g = 9.81
time_simulation = 2000
Qtur_init = Qtunnel = 50
Qtur_final = 20
time_control = 10.0
fr = 0.01
Ltunnel = 1000.0
Dtunnel  = 5.0
Asurgetank = 150.0
dt = 0.1

Qchangerate = (Qtur_final-Qtur_init)/(time_control)
Atunnel = 3.14*(Dtunnel**2)/4
v = Qtunnel/Atunnel
k1 = g*Atunnel/Ltunnel
k2 = fr*Ltunnel/(2*g*Dtunnel*Atunnel**2)

nt = int(time_simulation/dt)
time = np.zeros(nt)
tankwl = np.zeros(nt)
tankwl[0] = z1 = (-1.0)*k2*Qtur_init**2
Q1 = Qturbine1 = Qtur_init

for n in range (1,nt):
```

```
    time[n] = time[n-1] + dt
    if(time[n]<time_control):
        Qturbine2 = Qtur_init+Qchangerate*time[n]
    else:
        Qturbine2 = Qtur_final
    dQdt1 = (-1.0)*k1*(z1+k2*Q1*abs(Q1))
    dzdt1 = (Q1-Qturbine1)/Asurgetank
    Q2s = Q1+dQdt1*dt
    z2s = z1+dzdt1*dt
    dQdt2 = (-1.0)*k1*(z2s+k2*Q2s*abs(Q2s))
    dzdt2 = (Q2s-Qturbine2)/Asurgetank
    Q2 = Q1+dt*(dQdt1+dQdt2)/2
    z2 = z1+dt*(dzdt1+dzdt2)/2
    Q1 = Q2
    tankwl[n] = z1 = z2

fig = plt.figure()
ax = fig.add_subplot(1, 1, 1)
ax.plot(time, tankwl)
ax.set_xlabel('Time (s)')
ax.set_ylabel('Water surface elevation above datum (m)')
plt.show()
```

The variables used in this program are described below:

| Variable | Description | Variable | Description |
|---|---|---|---|
| g | Acceleration due to gravity (m/s²) | Ltunnel | Length of tunnel (m) |
| time_simulation | The time for simulation to run (s) | Dtunnel | Diameter of tunnel (m) |
| time_control | The time for controlling turbine flow | Asurgetank | Cross-sectional area of surge tank (m²) |
| Qtur_init, Qtunnel | Initial steady-state discharge in tunnel and turbine (m³/s) | dt | Incremental time step for carrying out computations (s) |
| fr | Darcy-Weisbach friction factor (-) | nt | Number of time points for plotting |

The above program in Python is run with the data specified in lines 5 to 14, which considers the flow to the turbine reducing from an initial steady-state value of 50 to 20 m³/s over a period of 10 s. Also, the initial discharge through the tunnel is the same as that flowing to the turbine. The surge tank is considered to be a cylindrical tower-like structure having a constant cross-sectional area, but may be changed if required, to one of non-uniform cross section. On running the code, the graphical output shown in Figure 3.12 is obtained. The datum for the level is taken as the initial

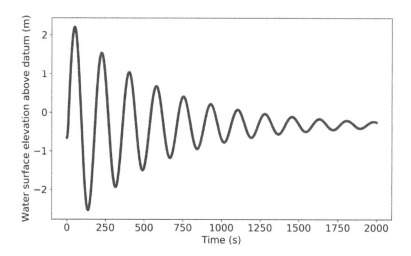

**FIGURE 3.12**   Variation of water level in surge tank with time. (Color image available in eBook).

steady-state elevation of the water surface in the surge tank. It is evident that the flow friction factor is responsible for the damping of the oscillations while the area of the surge tank controls the amplitude.

### 3.3.6   SECOND-ORDER ODE: STEADY-STATE GROUNDWATER TABLE PROFILE FOR RECHARGE AND WITHDRAWAL

The groundwater table in unconfined aquifers may be computed for given values of recharge and withdrawal using Equation (3.12), which is reproduced below:

$$\frac{K}{2}\frac{d^2\left(h^2\right)}{dx^2} = -R \tag{3.12}$$

The variable $R$ on the right hand side of Equation (3.12), represents the recharge rate due to infiltration of water from above. The same variable may also be used for specifying withdrawal through wells, simply by changing its sign. For solving Equation (3.12), the complete length of the one-dimensional domain is divided uniformly into discrete segments and a finite difference equation is written for each segment. Computational nodes mark the interface between two segments. There are also two boundary nodes at the two ends of the domain, where either the value of the dependent variable $h$ is known (the Dirichlet condition), or its slope is known (the Neumann condition). If the length of each segment is $\Delta x$, then the equivalent finite-difference equation for the $i^{th}$ computational node may be written as below:

$$\frac{\left(h^2\right)_{i-1} - 2\left(h^2\right)_i + \left(h^2\right)_{i+1}}{\left(\Delta x\right)^2} = -\left(\frac{2R_i}{K}\right) \tag{3.32}$$

In Equation (3.32), $R_i$ is the rate of recharge over the node $i$ and $K$, the hydraulic conductivity, is assumed to be constant for all the segments. Writing the equation for all the interior computational nodes, a system of equations is formed. Together with the values of $h$ known at the boundary nodes, which we assume to be specified in the present example demonstrated through the Python program given below, the system is solved for the unknown values of $h$ at the interior nodes. The form of the equation being non-linear, the method of Newton–Raphson is applied by first finding the vector of errors $\{\Delta h\}$ using the Jacobian matrix of the original system of equations evaluated from the derivatives of the unknown variables, as shown below:

$$
\begin{bmatrix}
1 & 0 & 0 & 0 & & & & \\
2h_1 & -4h_2 & 2h_3 & 0 & & & & \\
0 & \cdots & \cdots & \cdots & & & & \\
0 & 0 & 0 & 2h_{i-1} & -4h_i & 2h_{i+1} & 0 & \\
& & & & \cdots & \cdots & \cdots & \\
& & & & 0 & 2h_{N-2} & -4h_{N-1} & 2h_N \\
& & & & 0 & 0 & 0 & 1
\end{bmatrix}
\begin{Bmatrix}
\Delta h_1 \\
\Delta h_2 \\
\cdots \\
\Delta h_i \\
\cdots \\
\Delta h_{N-1} \\
\Delta h_N
\end{Bmatrix}
$$

$$
= \begin{Bmatrix}
h_1 - H_{high} \\
\cdots \\
\cdots \\
h_{i-1}^2 - 2h_i^2 + h_{i+1}^2 + (\Delta x)^2 \left(\dfrac{2R_i}{K}\right) \\
\cdots \\
\cdots \\
h_N - H_{low}
\end{Bmatrix}
\tag{3.33}
$$

Equation (3.33) is a tridiagonal matrix which may be solved with ease, especially with one of the available linear algebraic equation solvers in Python. Note that the calculations involve the evaluation of the elements of the Jacobian matrix from a known set of values of the variables $\{h\}$. Hence, for the computations to commence, a set of guessed values of $\{h\}$ is required initially. Once the error vector $\{\Delta h\}$ is found, it is used to update the values of the previous estimates of $\{h\}$ as follows:

$$
\begin{Bmatrix}
h_1 \\
h_2 \\
h_i \\
h_{N-1} \\
h_N
\end{Bmatrix}^{updated}
=
\begin{Bmatrix}
h_1 \\
h_2 \\
h_i \\
h_{N-1} \\
h_N
\end{Bmatrix}^{old}
-
\begin{Bmatrix}
\Delta h_1 \\
\Delta h_2 \\
\Delta h_i \\
\Delta h_{N-1} \\
\Delta h_N
\end{Bmatrix}
\tag{3.34}
$$

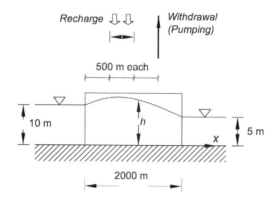

**FIGURE 3.13** One-dimensional seepage flow with distributed recharge and point withdrawal.

In the present example, we assume the values of $h$ at the boundary nodes ($H_{high}$ and $H_{low}$) to be known, whence in Equation (3.33), these values would not change. However, if the solution has to be modified for incorporating a Neumann condition at the boundary, it would be necessary to evaluate the values of $h$ at the boundary as well since these would not be known beforehand. For example, if there is a no-flow or an impervious boundary at the left end of the domain (that is, at $x = 0$), then we need to use a condition of the form $h_1 = h_2$, construing it to be an equivalent of the zero-gradient of $h$ at this location. In the example demonstrated through the following program in Python, both recharge and pumping are assumed to be present. The sample data used are shown in Figure 3.13.

The corresponding Python program is as follows:

```python
# 1D seepage flow with recharge and pumping
import numpy as np
import matplotlib.pylab as plt

length = 2000
npoints = 101
Hconductivity = 25
Hhigh = 10
Hlow = 5
errorallow = 0.001
maxiter = 10

x = np.linspace(0, length, npoints)
dx = x[1]-x[0]

R = np.zeros(npoints)
for i in range (25, 50):
    R[i] = 0.010
```

```
R[75] = -0.25

h = np.ones(npoints)*10.0

iter = 0
max_error = 1.0
while (max_error > errorallow):
    iter = iter+1
    if(iter > maxiter): break
    A = np.zeros((npoints,npoints))
    rhs = np.zeros(npoints)
    for i in range(1,npoints-1):
        A[i,i-1] = 2*h[i-1]
        A[i,i] = -4*h[i]
        A[i,i+1] = 2*h[i+1]
        rhs[i] = h[i-1]**2-2*h[i]**2+h[i+1]**2+R[i]*
  dx**2/Hconductivity

    A[0,0] = 1.0
    rhs[0] = h[0]-Hhigh
    A[npoints-1,npoints-1] = 1.0
    rhs[npoints-1] = h[npoints-1]-Hlow

    delh = np.linalg.solve(A,rhs)
    max_error = np.max(abs(delh))
    print("iter = ",iter," .... max_error = ",max_error)
    h -= delh

fig = plt.figure()
ax = fig.add_subplot(1, 1, 1)
ax.plot(x/1000,h,'o-')
ax.set_xlabel('Distance (km)')
ax.set_ylabel('Head (m)')
plt.show()
```

The variables used in this program are described below:

| Variable | Description | Variable | Description |
|---|---|---|---|
| length | Length of the one-dimensional domain (m) | Hconductivity | Hydraulic conductivity $(m\text{-}day^{-1})$ |
| npoints | Number of space points dividing the domain uniformly | errorallow | Error tolerance value for iterative computation of h (m) |

*(Continued)*

| Variable | Description | Variable | Description |
|---|---|---|---|
| Hhigh | Higher elevation of water above datum (m) | maxiter | Maximum number of iterations allowed |
| Hlow | Lower elevation of water above datum (m) | R | Array for specifying recharge or withdrawal (m-day$^{-1}$) |
| | | h | Array for storing h (m) [initial guesses or computed values] |

The graphical result obtained by running the above Python program is shown in Figure 3.14.

It may be seen that the distributed recharge spread between 0.5 km and 1 km is causing some amount of rise of the water table. However, the presence of the pumping well at 1.5 km from origin causes a deep depression of the water table.

### 3.3.7   SECOND-ORDER ODE: COMPUTING THE STEADY-STATE CONCENTRATION PROFILE FOR POINT LOADINGS IN ONE-DIMENSIONAL CHANNEL FLOW

Here, we discuss one more water quality problem similar to the determination of *BOD* and *DO* in stream, as shown in Section 3.3.4. Here too, we assume that effluents are being released in a one-dimensional channel (Figure 3.4), and we wish to determine the steady-state distribution of the concentration of a certain contaminant being released with the effluents along the length of the channel. There are a few differences, though, with the computation of the *BOD* and *DO* in the channel.

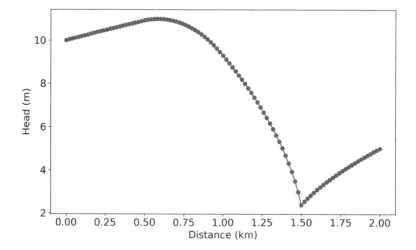

**FIGURE 3.14**  Plot of the groundwater table for one-dimensional seepage flow occurring under a distributed recharge and a point-well pumping. (Color image available in eBook).

Although the contaminant that we are assuming may be non-conservative (that is, it may decay or grow with time by reacting with the atmosphere or water), it is also subjected to dispersion. Note that the transport of the basic water quality parameters, that is the *BOD* and the *DO*, were defined by a coupled differential equation. However, for defining the transport and fate of a single constituent of a contaminant, it is normal to consider a single second-order differential equation of the form given by Equation (3.13), and shown again below:

$$E\frac{d^2C}{dx^2} - v\frac{dC}{dx} - kC + S = 0 \qquad (3.13)$$

In the above equation, $E$ is the dispersion coefficient [$L^2T^{-1}$]; $v$ is the velocity of the channel [$LT^{-1}$]; and $k$ is the decay rate constant of the constituent [$T^{-1}$]. The source term, $S$, and concentration, $C$, should be in compatible units. For example, if $S$ is in [$ML^{-3}T^{-1}$], then C must be in [$ML^{-3}$]. It is further assumed that one condition each is specified at the two ends of the one-dimensional channel through which the flow is taking place. These conditions may either be of the Dirichlet type, where the value of the concentration of the constituent is specified at the boundary, or of the Neumann type, in which the gradient or slope of the concentration of the pollutant is assumed to be known. We consider the 20-km-long channel behind the reservoir of a dam (Section 3.1.3), whose backwater curve profile along with the velocities at different points were found out using a Python program (Section 3.3.3). Recall that the velocities which were evaluated by the code were stored in a "comma separated variable" file by the name GVF.v.csv, which we shall use in the Python program shown below. In this section, we shall assume that pollutants are being released at more than one location in the channel and it is required to determine the complete distribution of its concentration throughout the channel.

As with the two-point boundary value problems, we use a finite difference method to approximate the governing equation (Equation 3.13). Note that this equation, though of second-order, is linear. Hence, iterative solutions are not required in this case as was done for the seepage-flow groundwater-table determination problem discussed in Section 3.3.6.

Equation (3.13) is reordered in the following way for making it adaptable for further derivations.

$$-E\frac{d^2C}{dx^2} + v\frac{dC}{dx} + kC = S \qquad (3.35)$$

Note that the source term, $S$, has been kept on the right hand side of the equation, while the other terms involving the unknowns are retained on the left. The resulting difference equation, centred around the computational node $i$, is as follows:

$$-\frac{E}{\Delta x}\left[\frac{C_{i+1} - C_i}{\Delta x} - \frac{C_i - C_{i-1}}{\Delta x}\right] + \frac{1}{\Delta x}\left[v_{i+1}\left(\frac{C_i + C_{i+1}}{2}\right) - v_i\left(\frac{C_{i-1} + C_i}{2}\right)\right] + kC_i = S_i \qquad (3.36)$$

The above equation may further be expressed in terms of the concentrations at the nodes, which are the unknowns requiring to be solved.

$$C_{i-1}\left[-0.5v_i - \frac{E}{\Delta x}\right] + C_i\left[0.5(v_{i+1} - v_i) + \frac{2E}{\Delta x} + k(\Delta x)\right] + C_{i+1}\left[0.5v_{i+1} - \frac{E}{\Delta x}\right] = S_i(\Delta x) \quad (3.37)$$

In Equation (3.37), the source term is non-zero corresponding to the computational nodes for which point loads exist. The equations, written for all the interior nodes, are solved together with the boundary conditions assumed to be as under:

1. Inflow boundary (at $x = 0$): Concentration is zero
2. Outflow boundary (at $x =$ Length of channel): Gradient of concentration is zero

The two conditions at the boundaries have been arbitrarily chosen, and may be changed according to the problem definition. The corresponding program in Python is written as follows:

```
# Pollutant transport (steady state) in backwater GVF
# behind a dam
import numpy as np
import matplotlib.pylab as plt

length = 20000
npoints = 101
Diffusivity = 100
DecayCoeff = 0.0001
S = np.zeros(npoints)
S[10] = 0.01; S[50] = 0.005

x = np.linspace(0, length, npoints)
dx = x[1]-x[0]
v = np.loadtxt('GVF.v.csv',delimiter=',')

A = np.zeros((npoints,npoints))
rhs = np.zeros(npoints)
for i in range(1,npoints-1):
    A[i,i-1] = (-v[i]/2-Diffusivity/dx)
    A[i,i] = ((0.5*(v[i+1]-v[i]))+2*Diffusivity/
    dx+DecayCoeff*dx)
    A[i,i+1] = (v[i+1]/2-Diffusivity/dx)
    rhs[i] = S[i]*dx

A[0,0] = 1.0
A[npoints-1,npoints-2] = 1.0
A[npoints-1,npoints-1] = -1.0

Conc = np.linalg.solve(A,rhs)
```

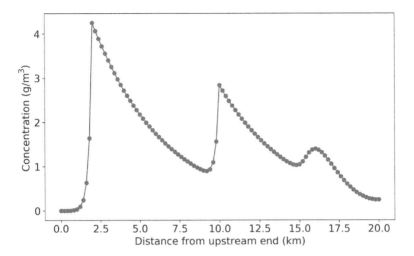

**FIGURE 3.15** Steady-state concentration profile of a pollutant in the backwater flow behind a dam. Concentrated point loads are imposed at 2 km and 10 km. Flow is from left to right. (Color image available in eBook).

```
fig = plt.figure()
ax = fig.add_subplot(1, 1, 1)
ax.plot(x/1000,Conc,'o-')
ax.set_xlabel('Distance from upstream end (km)')
ax.set_ylabel('Concentration (g/m3)')
plt.show()
```

The variables used in the program are described below:

| Variable | Description | Variable | Description |
|---|---|---|---|
| length | Length of the one-dimensional domain (m) | Diffusivity | Diffusivity coefficient $(m^2 s^{-1})$ |
| npoints | Number of space points dividing the domain uniformly | DecayCoeff | Decay rate constant or coefficient $(s^{-1})$ |
| S | Array for specifying point source at any given node (gram-$m^{-3}$) | Conc | Array for storing concentration values (gram-$m^{-3}$) |

Note that in the above program, the boundary conditions are imposed in the following ways:

1. Zero concentration at inflow boundary (at $x = 0$) is directly implemented by the statement $A[0,0] = 1.0$
2. Zero concentration gradient at outflow boundary (at $x =$ Length of channel) is implemented by equating the last unknown with the one before, that is, A[npoints-1,npoints-2] = 1.0 and A[npoints-1,npoints-1] = -1.0.

On running the above program for the given data specified in lines 6 to 11, a graphical output is obtained which shows the steady-state concentration profile of the pollutant along the length of the channel (Figure 3.15). Point-loading of pollutants have been specified through the $S$ array, which is initialized to zero at the beginning but populated with specific values corresponding to the locations of the pollutant sources. In the above program, two concentrated point sources of pollutants are specified as S[10] and S[50]. The length of the channel is 20 km and the two loading points are thus located at 2 km and 10 km, respectively, from the origin. Note that the program has used the velocities computed by the gradually varied flow code described in Section 3.3.3, which was run with a low discharge ($Q = 10$ m$^3$s$^{-1}$) and a weir height ($y_d$) equal to 5 m.

## REFERENCES

Benedini, M. and Tsakiris, G. (2013). *Water Quality Modelling for Rivers and Streams.* Springer. https://www.springer.com/gp/book/9789400755086

Bungay, H. R. (1998). *Environmental Systems Engineering.* Kluwer Academic Publishers, New York, USA.

Chapra, S. and Canale, R. (2021). *Numerical Methods for Engineers.* McGraw-Hill Education, 8th edition. https://www.mheducation.com/highered/product/numerical-methods-engineers-chapra-canale/M9781260232073.html

Chaudhry, M. H. (2008). *Open Channel Flow.* Springer, 2nd edition.

CPCB (2006).*Water quality status of Yamuna river (1999–2005).* Central Pollution Control Board, Delhi, India.

Thomann, R. V. and Mueller, J. A. (1987). *Principles of Surface Water Quality Modeling and Control.* Harper-Collins, New York. https://www.pearson.com/uk/educators/higher-education-educators/program/Thomann-Principles-of-Surface-Water-Quality-Modeling-and-Control/PGM409432.html

McCutcheon, S. C. (1989). *Water quality modeling.* Vol. 1, CRC Press, Boca Raton, FL, USA. https://www.routledge.com/Water-Quality-Modeling-River-Transport-and-Surface-Exchange-Volume-I/McCutcheon/p/book/9780849369711

Chaudhry, M. H. (2014). *Applied Hydraulic Transients.* Springer. 3rd edition. https://www.springer.com/gp/book/9781461485377

Chapra, S. C. (1997). *Surface Water Quality Modeling.* Waveland Press Inc. https://www.waveland.com/browse.php?t=378

Jaiswal, M., Hussain, J., Gupta, S. K., Nasr, M. and Nema, A. K. (2019) "Comprehensive evaluation of water quality status for entire stretch of Yamuna River, India", *Environ. Monit. Assess.*, Vol. 191, Article 208.

Sharma, D. and Singh, R. K. (2009). "DO-BOD modeling of River Yamuna for national capital territory, India using STREAM II, a 2D water quality model", *Environ. Monit. Assess.*, 159, 231–240.

Streeter, H. W. and Phelps, E. B. (1925) A Study of the pollution and natural purification of the Ohio river. III. Factors concerned in the phenomena of oxidation and reaeration, *U. S. Public Health Bulletin no. 146*, Reprinted by U.S. Department of Health, Education and Welfare, Public Health Service, 1958.

Wang, H. F. and Anderson, M. P. (1982). *Introduction to Groundwater Modeling: Finite Difference and Finite Element Methods.* Academic Press.

# 4 Partial Differential Equations in Surface Hydrology, Free Surface Flows, and Ideal Fluid Flows

In this chapter, we look into some representative problems from the field of surface hydrology and hydraulics related to runoff modelling over catchment surfaces, and free surface flows in open channels and shallow water basins. We also add an example of ideal-fluid flow in a rectangular duct. All these phenomena are described using partial differential equations (PDEs). The first of these deals with the generation of surface runoff by a known or given value of effective rainfall. Flows in open channels and rivers, or in natural and man-made water bodies like lakes or reservoirs, are classified as free surface flows, and are commonly encountered in the nature. In all such cases, the top surface of the flowing water is open to the atmosphere. In hydraulics, these are contrasted against flows through conduits or pipes where the difference of pressure between the two ends of a pipe drives the flow. In free surface flows, the movement of water is driven by gravity. The example on ideal-fluid flow may be used to model idealized flows in pipes and conduits, for certain specific conditions.

In the previous chapters, equations related to free surface flows were either of the non-linear algebraic form or ordinary differential equations (ODEs). The examples included in this chapter are governed by PDEs in two variables. For the unsteady flow problems, of which only the one-dimensional case is considered, there are two independent variables: one the space and the other the time. A case of steady-state flow in a shallow water body is also discussed where the governing equations are characterized as partial differential equations in two variables but with both variables representing lengths – each measured along one of the perpendicular coordinate axes. For solving the differential equations, we apply the finite difference techniques for all the cases since these are easier to handle and easily amenable for conversion into codes, which the beginners may find helpful. The advanced reader may, however, explore the methods of finite elements and, especially, the finite volumes, which are presently being used extensively in solving hydraulics and free surface flow problems.

DOI: 10.1201/9780429288579-4

## 4.1  GOVERNING EQUATIONS OF FREE SURFACE FLOW

Unsteady free surface flows occurring in canals, streams, and rivers are often classi-fied as one-dimensional since the width and depth of the flowing water are compara-tively smaller than the length measured along the direction of flow. In this case, the variation in the flow variables like depth and velocity are approximated to be nearly the same across the entire section of the channel. Two-dimensional flows are those which may either be depth-averaged or width-averaged. In the former, the velocities are considered to be varying horizontally, normal to the direction of gravity, but assumed to be approximated by a single value, averaged over its depth, at a given point. This assumption may be used for flows in shallow lakes or in estuaries, where the horizontal or lateral extent of the water body far exceeds its average depth. The width-averaged approximations of flows are made use of while dealing with the hydrodynamics of deep channels where there may not be much variation in the flow variables across the width although noticeable changes occur vertically, in the direc-tion of gravity. Examples for such flows may occur in certain reservoirs behind dams, where the river valley behind is like a narrow gorge with steep sides. Finally, the complete three-dimensional case is the most general flow situation which may be of interest in studying, for example, the hydraulics in the vicinity of the intake to an outlet conduit from a reservoir. The velocities in the reservoir would generally be varying in all three directions close to the intake.

In this chapter, and indeed in this book, we limit ourselves to the most common free surface hydraulic conditions – the one-dimensional flows in channels and depth-averaged flows in shallow water bodies. In the former, the bed or the bottom of the channel is assumed to be known and so is the bottom elevation of the shallow water basin in the second case. In a channel, the bed is sometimes assumed to be sloping uniformly in one direction, which makes the calculations simpler. Similarly, the cal-culations in a two-dimensional shallow lake would be simplified if the bottom topog-raphy, generally known as bathymetry, is assumed to be either completely flat or sloping uniformly in one direction. The surface runoff examples presented herein are also free surface flows, although simplified equations are used in modelling them.

### 4.1.1  GOVERNING EQUATIONS OF FLOW IN A PRISMATIC CHANNEL

For a flow reach between two sections of a prismatic channel, as shown in Figure 4.1, the governing equations are derived from the following considerations:

1. The continuity equation, derived by considering the net inflows, outflows, and storage within the section, and
2. The momentum equation, considering the net forces and momentum in the flow direction.

Although a more general equation may be available for a non-prismatic channel, the prismatic channel equation is presented here since all the examples dealt with in the following sections conform to this assumption. Following the derivations given in standard references like Chow et al. (1988) or Chaudhry (2008), we may write the

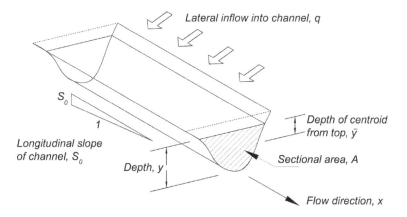

Lateral inflow into channel, q

$S_0$

1

Longitudinal slope
of channel, $S_0$

Depth, y

Depth of centroid
from top, $\bar{y}$

Sectional area, A

Flow direction, x

**FIGURE 4.1** Section through a prismatic one-dimensional channel.

equations of continuity and momentum, referred to as the St. Venant equations of flow, by the following pair of equations:

$$B\frac{\partial y}{\partial t} + \frac{\partial Q}{\partial x} = q \tag{4.1}$$

$$\frac{\partial Q}{\partial t} + \frac{\partial(Qu)}{\partial x} + gA\frac{\partial y}{\partial x} = V_x q + gA\left(S_0 - S_f\right) \tag{4.2}$$

In both Equations (4.1) and (4.2), $q$ is the lateral inflow entering the channel from its sides, with a velocity component $V_x$ in the direction of channel flow. Of the other variables, $Q$ is the discharge passing through the section, $y$ is the depth of flow, $u$ is the velocity of flow in the channel, $A$ is the cross-sectional area, and $S_0$ and $S_f$ are the channel bed slope and friction slope, respectively. The independent variables are $x$, defined as the coordinate along the flow direction, and $t$, the time. We shall ignore the term $V_x$ in the subsequent applications of the above pair of equations, since it is usually negligible. The equations are said to be derived in the, so-called, conservative form. Oftentimes, the non-conservative forms of the continuity and momentum equations are used, which are expressed as follows:

$$u\frac{\partial y}{\partial x} + y\frac{\partial u}{\partial x} + \frac{\partial y}{\partial t} = 0 \tag{4.3}$$

$$\frac{\partial u}{\partial t} + u\frac{\partial u}{\partial x} + g\frac{\partial y}{\partial x} - g\left(S_0 - S_f\right) = 0 \tag{4.4}$$

Equation (4.4), the equation for momentum in the flow direction, is known as the dynamic wave equation. One of the simplified forms of this equation is used in certain cases, especially in channels with a significant bed slope. Called the kinematic wave equation, it is obtained by neglecting the differentials and retaining only the two slope terms, $S_0$ and $S_f$. This form of the momentum equation assumes that the downward force due to gravity balances the opposing force of friction. Hence, the acceleration of the water is neglected, both in time or space, although this form of the equation is also used sometimes to roughly model unsteady flows in steep channels. Although the results may not be perfect, it may still be possible to obtain quick solutions to, say, flood wave movement in steep mountainous channels where the inherent assumptions in applying the kinematic wave approximation may not yield too different a result from those using the complete dynamic equations.

## 4.1.2   IDEAL FLUID FLOW

The ideal fluid flow concept may be used to model irrotational fluid flows, also called the potential flows. The PDEs governing such phenomena are given by a Laplace-type equation of the following form:

$$\frac{\partial^2 \psi}{\partial x^2} + \frac{\partial^2 \psi}{\partial y^2} = 0 \tag{4.5}$$

In the above equation, $\psi$ stands for the stream function from which the velocities in the two coordinate directions, $x$ and $y$, may be found out using the two equations given below:

$$u = \frac{\partial \psi}{\partial y} \tag{4.6a}$$

and

$$v = -\frac{\partial \psi}{\partial y} \tag{4.6b}$$

## 4.1.3   GOVERNING EQUATIONS OF TWO-DIMENSIONAL DEPTH-AVERAGED FLOWS

The governing equations for flows that may be approximated as being two-dimensional in the horizontal plane, comprises of three independent equations. The first of these is derived from the considerations of mass balance or continuity, while the other two are obtained using the balance of forces and momenta in the two horizontal directions. The examples given in the previous sections could be classified as partial differential equations (PDEs) in two variables, while those for the two-dimensional depth-average flows considered here are PDEs in three variables. In the Cartesian rectangular coordinates, the continuity equation may be written as follows (Chaudhry, 2008):

$$\frac{\partial h}{\partial t} + \frac{\partial (uh)}{\partial x} + \frac{\partial (vh)}{\partial y} = 0 \tag{4.7}$$

In Equation (4.7), $u$ and $v$ are the values of the velocities in $x$- and $y$-directions, respectively, averaged over the depth of flow, $h$. If there is a source of flow entering the domain, for example, as rainfall with an intensity $i$, then it may be included on the right hand side of Equation (4.7). The momentum equations are given as:

$$\frac{\partial u}{\partial t} + u\frac{\partial u}{\partial x} + v\frac{\partial u}{\partial y} = -g\frac{\partial h}{\partial x} + gn^2\frac{u\sqrt{u^2+v^2}}{h^{4/3}} \tag{4.8}$$

$$\frac{\partial v}{\partial t} + u\frac{\partial v}{\partial x} + v\frac{\partial v}{\partial y} = -g\frac{\partial h}{\partial y} + gn^2\frac{v\sqrt{u^2+v^2}}{h^{4/3}} \tag{4.9}$$

## 4.2 NUMERICAL METHODS FOR SOLVING THE FLOW EQUATIONS

The examples and corresponding equations for the one- or two-dimensional flows given in the previous sections are of two types; those representing the free surface flows are non-linear first-order PDEs, while that for the ideal fluid flow is a linear second-order PDE. However, for both types of equations, the finite difference technique is used here for solving them numerically. We choose the explicit schemes for most of the cases, except while solving the propagation of a surge wave in a trapezoidal channel, where the Preissmann implicit scheme is employed (Chaudhry, 2008).

### 4.2.1 SOLVING THE KINEMATIC WAVE EQUATION FOR FLOW IN A PRISMATIC CHANNEL WITH LATERAL INFLOWS

For flows taking place in a prismatic channel and governed by the kinematic wave approximation of the St. Venant equations of flow, the following pair of equations are solved together (Chow et al., 1988):

$$\frac{\partial Q}{\partial x} + \frac{\partial A}{\partial t} = q \tag{4.10}$$

$$S_0 = S_f \tag{4.11}$$

In the equations above, the terms are as discussed before, with $A$ being the cross-sectional area of flow in the channel. The friction slope, $S_f$, may be expressed using the Manning's (or the Chezy) equation as below:

$$Q = \frac{S_0^{1/2}}{nP^{2/3}}A^{5/3} \tag{4.12}$$

Here, $P$ is the wetted perimeter and $n$ is the Manning's channel friction coefficient and $S_0$ is the longitudinal slope of the channel bed. Expressing the area of flow, $A$, in terms of the other variables in Equation (4.12), we may write:

$$A = \left( \frac{nP^{2/3}}{S_0^{1/2}} \right)^{3/5} Q^{3/5} \tag{4.13}$$

Further, since the expression $\dfrac{\partial A}{\partial t}$ in Equation (4.10) may be written as $\dfrac{\partial A}{\partial Q} \cdot \dfrac{\partial Q}{\partial t}$, and taking a derivative of the variable $A$ with respect to $Q$ from Equation (4.13), we obtain the following single equation, after appropriate substitution, as:

$$\frac{\partial Q}{\partial x} + \alpha \beta Q^{\beta-1} \frac{\partial Q}{\partial t} = q \tag{4.14}$$

In Equation (4.14), the only dependent variable is the discharge, $Q$, while there are two independent variables, $x$ and $t$. The coefficients $\alpha$ and $\beta$ in the second term are equal to $\left( \dfrac{nb^{2/3}}{S_0^{1/2}} \right)^{3/5}$ and 3/5 or 0.6, respectively, for a rectangular section of width $b$.

For computing the discharge along the length of the channel, the finite difference numerical method is applied to Equation (4.14), which requires the length of the channel to be divided equally into a given number of segments. The discharges at the computational nodes, assumed to be located at the interface of neighbouring segments, are computed at specified intervals of time. At the upstream boundary node of the channel, the depth of flow is assumed to be known at all times. The approximations of the derivative of the discharge $Q$ with the space variable $x$ for the $i^{th}$ computational node may be expressed by Equation (4.15), given below. The subscript $i$ in this equation is the node at which the discharge is being computed while $i-1$ is one space point behind, that is the node just upstream of node $i$.

$$\frac{\partial Q}{\partial x} = \frac{Q_i^{j+1} - Q_{i-1}^{j+1}}{\Delta x} \tag{4.15}$$

In Equation (4.15), the terms in the numerator on the right stand for the discharges at nodes $i$ and $i-1$ at $j+1$ time level, respectively. The term in the denominator, $\Delta x$, stands for the length of a segment. The derivative in time for the discharge variable, $Q$, at the $i^{th}$ node is computed as in Equation (4.16), where the current time level is designated as $j$ and the future time level as $j+1$, separated by a time interval of $\Delta t$. It is assumed that the values of the variables at the (current) $j^{th}$ time level are known, and those at the next time level are to be found out. Thus:

$$\frac{\partial Q}{\partial t} \approx \frac{Q_i^{j+1} - Q_i^{j}}{\Delta t} \tag{4.16}$$

In Equation (4.16), the terms in the numerator on the right stand for the discharges at node $i$ at $j+1$ and $j$ time levels, respectively. We may plug in the space and time derivative approximations from Equations (4.15) and (4.16) into Equation (4.14), but we still need to approximate a value for the discharge variable $Q$ appearing in the second term on the left in Equation (4.14). For this, we may implement the following averaging process of two values of discharge in time and space as given by Chow et al. (1988):

$$Q \approx \frac{Q_i^j + Q_{i-1}^{j+1}}{2} \qquad (4.17)$$

The lateral inflow, $q$, appearing on the right of Equation (4.14) is assumed to remain constant over time and space. Substituting the above approximations into Equations (4.14), we obtain Equation (4.18), given below, for the unknown discharge $Q_i^{j+1}$ at time level $j+1$ and space point $i$ in terms of other variables. Beginning with known values at an assumed zero time level, the computations proceed for each subsequent time level and for all computational nodes, except the upstream-most node, where the values of all variables are assumed to be known at all times. This process is continued until a specified final time is reached.

$$Q_i^{j+1} = \frac{\frac{\Delta t}{\Delta x} Q_{i-1}^{j+1} + \alpha\beta\left(\frac{Q_i^j + Q_{i-1}^{j+1}}{2}\right)^{\beta-1} Q_i^j + \Delta t \cdot q}{\frac{\Delta t}{\Delta x} + \alpha\beta\left(\frac{Q_i^j + Q_{i-1}^{j+1}}{2}\right)^{\beta-1}} \qquad (4.18)$$

## 4.2.2 Routing a Flood Wave by the Kinematic Wave Approximation in a Triangular Channel

In this section, we compute the variations in flow along a steep triangular channel, through which a flood wave is assumed to pass. Under the given condition of a steep slope, the kinematic wave approximation of the St. Venant equations is assumed to be valid for routing the flood flow and finding the resulting discharge at the channel's outlet end. For solving the problem using the finite difference method, the channel length is divided into segments, separated by computational nodes. By substituting Equation (4.13) into Equation (4.11), we obtain Equation (4.19) for the unknown area variable, $A$, at the time level $j+1$ and space point $i$ in terms of the area variable at neighbouring computational nodes. Unlike Equation (4.18), which was obtained in terms of the discharge $Q$ for the prismatic channel discussed in Section 4.1.1, Equation (4.19) is used to find out the area, $A$, at all computational nodes and at all times, starting from an initial or given time. At the upstream-most node, the flood hydrograph is assumed to be known. For approximating the derivatives of the variable $A$ in space and time, expressions similar to those used for $Q$, given by Equations (4.15) and (4.16), are made use of. For approximating the area $A$ itself, an expression similar to that of Equation (4.17) is employed. The resulting equation obtained is as follows:

$$A_i^{j+1} = \frac{A_i^j + \left(\frac{\alpha\beta\Delta t}{\Delta x}\right)\left(\frac{A_i^j + A_{i-1}^{j+1}}{2}\right)^{\beta-1} A_{i-1}^{j+1}}{\left(\frac{\alpha\beta\Delta t}{\Delta x}\right)\left(\frac{A_i^j + A_{i-1}^{j+1}}{2}\right)^{\beta-1}} \qquad (4.19)$$

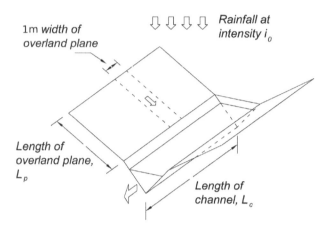

**FIGURE 4.2**   An open-book configuration of an idealized catchment (after Wooding, 1965).

### 4.2.3   OPEN-BOOK CATCHMENT HYDROGRAPH WITH THE KINEMATIC WAVE APPROXIMATION

An early proposition of a hydraulic model for a simplified geometry of a rectangular catchment was the "open-book" configuration of Wooding (1965), as shown in Figure 4.2. In this model, it is assumed that there are two "leaves" of the catchment spread open as an open book with a central channel, shown here to be triangular although a more general shape may be considered. A steady uniform rainfall is assumed to fall over the catchment at a rate $i_0$ [LT$^{-1}$]. Each half plane of the catchment may be assumed to be of different lengths, though we assume the same length $L_p$ for both in the present case. The channel is assumed to have a length $L_c$. Similarly, the slopes of the catchment planes and the channel may be different, and are designated here as $S_p$ and $S_c$, respectively. The rain falling over the catchment planes may be thought of as lateral inflow to a representative strip of the catchment (which may be of unit width, for the ease of calculations), as shown in Figure 4.2. In turn, the runoffs generated by the rain falling over the combination of the catchment planes contribute to the central channel as lateral flows from either side.

We apply the kinematic wave model twice in this case; once for computing the runoff generated from the unit-strip on the catchment, and then for estimating the outflow from the end of the channel. The equations are the same as used in the previous sections, that is, the flow emerging out of a catchment strip is solved using the flow calculation steps for a rectangular channel (Section 4.2.1) and that of the triangular channel solved using the procedure outlined for flood routing in Section 4.2.2.

### 4.2.4   SIMULATION OF UNSTEADY FLOWS IN A CHANNEL USING THE ST. VENANT EQUATIONS

A surge wave is a dynamic phenomenon of interest in practical hydraulic engineering which may be generated, among others, in hydropower canals when the downstream flow-control gate is closed or opened in a relatively short time. It may also occur in some estuaries when an extreme rise of the sea water level during the tides forces a bulk of water rushing into the river, causing a surge wave. Although the depths behind a surge wave may be analysed by assuming a pseudo-steady condition, as discussed in Section 2.1.2, in order to know the position or velocity of the surge wave in time, one has to solve the complete dynamic equations of flow (the St. Venant equations, Equations 4.1 and 4.2) with appropriate boundary conditions. Although there are many numerical methods to solve the equations for such a case, we chose the implicit finite difference scheme of Preissmann (Chaudhry, 2008) to demonstrate the implementation of the numerical computation schemes. Of course, other methods, especially the explicit finite difference methods, or the finite element and finite volume techniques may also be used alternatively in this case. Following Chaudhry (2008), we need to first divide the channel length into equal segments, with computation nodes in between, similar to those for the other problems in the previous sections. Denoting the computation nodes by the index $i$, and the time level by $k$, we may write the Preissmann scheme discretization for the governing equations as in the pair of equations given below:

The continuity equation:

$$\left[\left(A_i^{k+1} + A_{i+1}^{k+1}\right) - \left(A_i^k + A_{i+1}^k\right)\right] + 2\left(\frac{\Delta t}{\Delta x}\right)$$
$$\left\{\theta\left[(VA)_{i+1}^{k+1} - (VA)_i^{k+1}\right] + (1-\theta)\left[(VA)_{i+1}^k - (VA)_i^k\right]\right\} = 0 \qquad (4.20)$$

The momentum equation:

$$\left\{\begin{array}{l}\left[(VA)_i^{k+1} + (VA)_{i+1}^{k+1}\right] - \\ \left[(VA)_i^k + (VA)_{i+1}^k\right]\end{array}\right\} + 2\left(\frac{\Delta t}{\Delta x}\right)\left\{\begin{array}{l}\theta\left[\left(V^2A + gA\bar{y}\right)_{i+1}^{k+1} - \left(V^2A + gA\bar{y}\right)_i^{k+1}\right] + \\ (1-\theta)\left[\left(V^2A + gA\bar{y}\right)_{i+1}^k - \left(V^2A + gA\bar{y}\right)_i^k\right]\end{array}\right\} \qquad (4.21)$$
$$-gA\Delta t\left\{\theta\left[\left(S_0 - S_f\right)_{i+1}^{k+1} - \left(S_0 - S_f\right)_i^{k+1}\right] + (1-\theta)\left[\left(S_0 - S_f\right)_{i+1}^k - \left(S_0 - S_f\right)_i^k\right]\right\} = 0$$

In Equations (4.20) and (4.21), most of the terms are explained earlier in this chapter. However, a few additional terms are used, for example, $\bar{y}$, which is the distance of the centroid below the free surface of the flow section; $\Delta t$ is the computational time step size; $\Delta x$ is the distance separating the computational nodes along the linear domain; and $\theta$ is a time weighting factor which may vary from 0, making the scheme completely explicit, to 1, which makes the scheme completely implicit. If a fractional value of $\theta$ is chosen, the scheme becomes semi-implicit. Writing the above pair of equations for each computational node and the two boundary node points at either end of the computational domain, a set of non-linear algebraic equations is generated. The method of Newton–Raphson (N-R) for solving a set of such equations is given in Section 2.3.3, following which an iterative computational scheme may be set up to solve for the flow variables at every time step progressively. The computations commence from specified values of the variables at all the nodes along the channel, assumed to be known at an initial or zero time level. At the two boundary nodes located at either extreme ends of the channel, however, one must also specify the conditions in terms of the flow variables for the entire duration of time for which the computations are required to be carried out. Hence, for the case of a surge phenomenon to be simulated in a channel, we may specify a constant discharge from the upstream of the channel; whilst at the downstream boundary, an existing initial velocity of flow may instantly be reduced to zero to recreate the situation of a suddenly closed flow-control gate at that location.

## 4.2.5 IDEAL FLUID FLOW EQUATION SOLVING

Although the example presented in this section is related to the study of an ideal, irrotational fluid flow (Figure 4.3), sometimes by its application, an approximate solution to practical problems may be obtained in order to have a fair idea of the actual hydraulics occurring in certain real life situations. The governing stream-function equation (Equation 4.8) may be solved by approximating the equations with a central difference scheme by discretizing the computational domain with a rectangular grid of size $\Delta x$ and $\Delta y$ in the x- and y-directions, respectively, as shown in Figure 4.4 for a typical problem of flow through a contraction of a flow conduit. Polar grids or boundary fitted orthogonal grids may be used under appropriate conditions, though only rectangular grids are used here for simplicity. Further, we assume that the grid spacing $\Delta x$ and $\Delta y$ are equal in both directions.

Figure 4.3 also shows the boundary conditions, in terms of the stream function $\psi$, which needs to be appropriately implemented in order to obtain a correct solution. Figure 4.4a shows a typical grid that may be constructed for solving the problem within the domain in Figure 4.3. Approximations to the governing equation (Equation 4.8) may be done following the typical computational cell with four surrounding neighbouring nodes (Vreugdenhill, 1989), as shown in Figure 4.4b. The boundary nodes, marked with dark circles in Figure 4.5, are not included in the computation

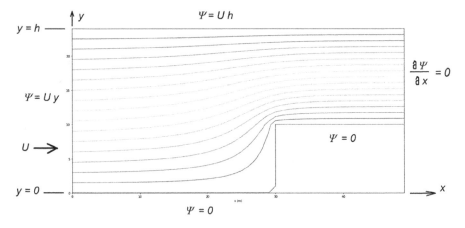

**FIGURE 4.3**  Ideal fluid flow through a sudden contraction (adapted from Vreugdenhil, 1989). (Color image available in eBook).

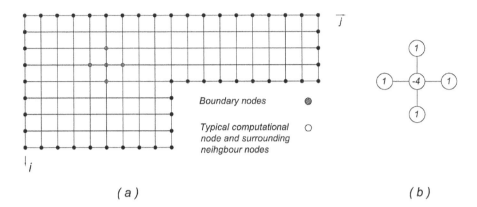

**FIGURE 4.4** (a) Computational grid for the problem of ideal fluid flow in a contraction of a conduit and (b) central computational node and neighbouring nodes of a computational cell with weighting coefficients.

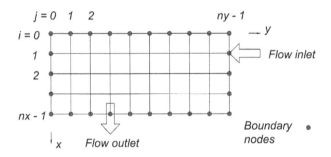

**FIGURE 4.5** Plan view of a shallow water domain in the form of a rectangular basin showing inflow and outflow points and an overlying computational grid.

scheme as the boundary conditions are imposed at these nodes. The interior nodes are all computational nodes and the calculations are performed for these nodes to find the values of the unknown stream-function variable $\psi$.

Considering the spacing of the grids to be $a$ in each direction, and denoting the grid indices as shown in Figure 4.4(a) by $i$ and $j$, respectively, in two perpendicular directions, we may discretize the governing equation (Equation 4.5) as given below:

$$\psi_{i+1,j} + \psi_{i-1,j} + \psi_{i,j+1} + \psi_{i,j-1} - 4\psi_{i,j} = 0 \tag{4.22}$$

Equation (4.22) is a linear combination of the variables of 5 neighbouring nodes and such equations may be developed for all the interior computational nodes, except those near the right-most boundary shown at the outlet end of the domain in Figure 4.4(a). For these nodes, a gradient of the stream-function variable is described as shown in Figure 4.3. Since the gradient is specified to be zero, and there is no acceleration or deceleration of the flow particles at this location, the following equation is written, approximating the condition.

$$\psi_{i,j-1} - \psi_{i,j} = 0 \qquad (4.23)$$

By writing the equations for all the boundary nodes at the right boundary, and combining them with those for the other nodes of the domain, a complete set of equations is obtained, which contains the exact number of unknowns as there are equations. Since the set of equations contains terms having only linear combination of the variables, a simple linear equation solver may be used to compute the unknowns at the interior computational nodes.

### 4.2.6 SIMULATION OF TWO-DIMENSIONAL DEPTH-AVERAGED FLOWS IN A SHALLOW BASIN

A simple rectangular shallow water body is considered for this example, such as a confined basin or a lake, the plan view of which with a grid superimposed for computations, is shown in Figure 4.5. It is also assumed that the bottom of the basin is horizontal. The $x$ and $y$ rectangular coordinate directions may be chosen as in the figure, or in any way that is convenient. The grids are assumed to be spaced equally in the $x$- and $y$-directions and the grid cell size may be denoted by $\Delta x$ and $\Delta y$ in either direction. The grid points in the $x$- and $y$-directions are numbered by the indices $i$ and $j$, respectively, and are counted from 0 onward, since it would be of help while converting it into a program in Python as the default counting for this language begins at 0. The physical boundary of the domain is marked by filled circles located at the intersection of the boundary grids. One or more inlet points may be assumed for defining the flows entering the basin. At all the inlet points, known flow rates or discharges should be known at all times for solving the governing equations. An outlet point is also assumed along the boundary of the domain where the water level is specified at all times.

For computing the velocities at the discretized grid points of the computational domain, the governing equations (Equations 4.7 to 4.9) are expressed in the following form (Chaudhry, 2008):

$$\begin{bmatrix} h \\ (hu) \\ (hv) \end{bmatrix}_t + \begin{bmatrix} (uh) \\ hu^2 + \left(\frac{1}{2}\right)gh^2 \\ (huv) \end{bmatrix}_x + \begin{bmatrix} (vh) \\ (huv) \\ hv^2 + \left(\frac{1}{2}\right)gh^2 \end{bmatrix}_y = \begin{bmatrix} q \\ gh\left(S_{0x} - S_{fx}\right) \\ gh\left(S_{0y} - S_{fy}\right) \end{bmatrix} \qquad (4.24)$$

Equation (4.24) is known as the "conservative form" as it expresses mathematically the conservation of mass, $x$-momentum and $y$-momentum, respectively, in its three rows. Note that the derivatives in time and two space directions ($x$ and $y$) are denoted by the subscript along each column of Equation (4.24). Further, the elements appearing on the right hand column of the above system of equations may be called the "source term" as these quantities are responsible for initiating the changes to the variables $h$, $u$, and $v$ appearing in the equations. Since the first row of Equation (4.24) represents the conservation of mass, the corresponding source term, $q$, would be the rate of flow per unit width, if there is any. Hence, when used in numerical computations, this term would be set to zero at all grid points except where the inflows are specified. Similarly, the source term for the last two equations representing the gravity components and friction forces would be zero if these are neglected, or specified according to the data of the given problem. Note that $S_{0x} = \left(\frac{\partial z}{\partial x}\right)$ and $S_{0y} = \left(\frac{\partial z}{\partial y}\right)$ are the slopes of the bed in the $x$- and $y$-directions, where $z$ is the bottom elevation of the basin above a datum. Similarly, $S_{fx}$ and $S_{fy}$ are the friction slopes, given by the following expressions:

$$S_{fx} = n^2 \frac{u\sqrt{u^2 + v^2}}{h^{1.33}} \tag{4.25a}$$

$$S_{fy} = n^2 \frac{v\sqrt{u^2 + v^2}}{h^{1.33}} \tag{4.25b}$$

We may write Equation (4.24) in a compact form as:

$$U_t + F_x + G_y = S \tag{4.26}$$

From here, we may apply any suitable numerical scheme for solving the set of equations given by Equation (4.26). However, for the present work, we shall use the finite difference methodology and employ the technique of Lax–Wendroff, which is a two-step algorithm centred in time, and centred in space step in the two dimensions (Lax and Wendroff, 1960). The discretization of the equations in time and space may be written for the first "half-step" as:

$$U_{i+1/2,j}^{n+1/2} = 0.5\left(U_{i+1,j}^n + U_{i,j}^n\right) - \frac{\Delta t}{\Delta x}\left(F_{i+1,j}^n + F_{i,j}^n\right) \tag{4.27a}$$

$$U_{i,j+1/2}^{n+1/2} = 0.5\left(U_{i,j+1}^n + U_{i,j}^n\right) - \frac{\Delta t}{\Delta y}\left(G_{i+1,j}^n + G_{i,j}^n\right) \tag{4.27b}$$

This is then followed by the "whole step", given as:

$$U_{i,j}^{n+1} = U_{i,j}^n - \frac{\Delta t}{\Delta x}\left(F_{i+1/2,j}^{n+1/2} + F_{i-1/2,j}^{n+1/2}\right) - \frac{\Delta t}{\Delta x}\left(G_{i,j+1/2}^{n+1/2} + G_{i,j-1/2}^{n+1/2}\right) \tag{4.28}$$

It may be kept in mind that initial conditions in the form of specified depth and velocities in the two directions need to be defined at all grid points to commence the computations. The boundary conditions, too, may need to be specified correctly for obtaining accurate results. For example, these may be defined as below for the $x$-boundary of the given problem as:

$$h_{\text{at }x\text{-boundary}} = h_{\text{interior neighbour}}, u_{\text{at }x\text{-boundary}} = 0, \text{and } v_{\text{at }x\text{-boundary}} = v_{\text{interior neighbour}}$$

Similarly at the $y$-boundary, the following may be defined:

$$h_{\text{at } y\text{-boundary}} = h_{\text{interior neighbour}}, u_{\text{at } y\text{-boundary}} = u_{\text{interior neighbour}}, \text{and } v_{\text{at } y\text{-boundary}} = 0$$

## 4.3   PYTHON PROGRAMS

The algorithms discussed in the previous sections are translated here into working Python language codes as given below. Note that all the problems discussed are governed by PDEs and the method of finite difference numerical technique is used exclusively for finding their solutions.

### 4.3.1   FLOW IN A RECTANGULAR CHANNEL WITH LATERAL INFLOWS SOLVED BY THE KINEMATIC WAVE EQUATION

The program for computing the flow taking place through a prismatic rectangular channel with lateral inflows using Equation (4.18) is presented below.

```
# Kinematic wave approximation for finding outflow from a
# rectangular channel
import numpy as np
import matplotlib.pylab as plt

b0 = 20.0
s0 = 0.01
mn = 0.035
lateralflow = 0.1
length = 100
nx = 10
timerainfall = 200
timesimulation = 400
dt = 1
nt = int(timesimulation/dt)

x = np.linspace(0, length, nx)
dx = x[1] - x[0]
Qold = np.ones(nx)*0.001
Qnew = np.ones(nx)*0.001
alpha = (mn*b0**0.6667/np.sqrt(s0))**0.6
beeta = 0.6
b1 = beeta-1
time = np.zeros(nt)
Qmidpoint = np.zeros(nt)
Qoutlet = np.zeros(nt)
midpoint = int(nx/2)

# Main program
for n in range(1,nt):
    time[n] = time[n-1] + dt
    print("time = ",time[n])
```

```
    dtdx = dt/dx
    if (time[n] < timerainfall):
        q = lateralflow
    else:
        q = 0.0

#   Computation for Q-s at all point for given time
    for i in range(1, nx):
        Qavg = 0.5*(Qold[i]+Qnew[i-1])
        numer = dtdx*Qnew[i-1]+(alpha*beeta*Qold[i]*(Qavg**b1))
    +dt*q
        denom = dtdx+(alpha*beeta*(Qavg**b1))
        Qnew[i] = numer/denom

    Qmidpoint[n] = Qnew[midpoint]
    Qoutlet[n] = Qnew[nx-1]
    Qold = Qnew

fig = plt.figure()
ax = fig.add_subplot(1, 1, 1)
ax.plot(time, Qmidpoint,label='At channel midpoint')
ax.plot(time, Qoutlet,label='At channel outlet')
ax.set_xlabel('Time (s)')
ax.set_ylabel('Discharge (m3s-1)')
plt.legend()
plt.show()
```

This program makes use of Equation (4.13), and by assuming that the wetted perimeter (P) is equal to the width of the rectangular section, simplifies the calculations for such channels. An element of error thus remains, as P should indeed be the width plus twice the flow depth.

The variables used in this program are described below:

| Variable | Description | Variable | Description |
|---|---|---|---|
| b0 | Width of the rectangular channel (m) | lateralflow | Lateral flow entering the channel $(m^2/s)$ |
| s0 | Longitudinal slope of the channel (-) | length | Length of channel (m) |
| mn | Manning's channel roughness coefficient $(m^{-1/3}s)$ | friction | Array storing the friction factors of the pipes |
| dt | Computational time step (s) | timerainfall | Time up to which rainfall occurs (s) |
| nx | Number of points into which the channel length is divided for computations | timesimulation | Total time up to which simulation is carried up to (s) |

On running the above Python code, the graphs plotting time versus discharge at two locations in the channel, one at its midpoint and the other at its outlet, are presented in Figure 4.6(a) for a lateral inflow of 0.1 $m^2s^{-1}$. The input data of lateral inflow in the program, when increased 10 times, and the resulting discharges at the two points are plotted in Figure 4.6(b). Although in either case, the total discharge at the mid-length of the channel is half that at the outlet, as expected, the time to reach a saturation discharge is seen to clearly vary with the intensity of lateral flows. A larger intensity causes the saturation to reach faster than that for a smaller lateral discharge. Interestingly, the results simulated by the Python program is equivalent to that of a rectangular plane receiving rainfall at a given intensity, which may be contrasted with the linearity assumptions in the application of the method of the unit hydrograph in surface hydrology.

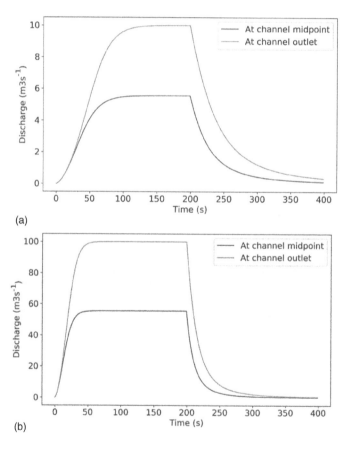

**FIGURE 4.6**   Discharges in a rectangular channel with (a) lateral flow of 0.1 $m^2s^{-1}$ and (b) lateral flow of 1.0 $m^2s^{-1}$. (Color image available in eBook).

## 4.3.2 Routing a Flood Hydrograph by the Kinematic Wave Approximation in a Triangular Channel

The problem discussed in Section 4.2.2 of a triangular channel with an incoming flood wave entering at its upstream end, is coded in Python. The kinematic wave approximation is used in this case, as was done in the preceding section, but the solution methodology is that given by Equation (4.19).

```python
# Kinematic wave approximation for flood routing through a
# triangular channel
import numpy as np
import matplotlib.pylab as plt

z = 2.0
s0 = 0.01
mn = 0.035
dt = 1
length = 100
nx = 10
Qinitial = 10.0
Qpeak = 200.0
Qfinal = 10.0
time_floodpeak = 20
time_floodend = 50
time_simulation = 100

nt = int(time_simulation/dt)
x = np.linspace(0, length, nx)
dx = x[1] - x[0]
zz=(z/(1+z*z))**0.3333
alpha = np.sqrt(s0)*zz/mn
beeta = 1.3333
b1 = beeta-1
Ainitial = (Qinitial/alpha)**(1/beeta)
Aold = np.ones(nx)*Ainitial
Anew = np.ones(nx)*Ainitial
time = np.zeros(nt)
Qinlet = np.ones(nt)*Qinitial
Qoutlet = Qinlet
dtdx = dt/dx

# Defining the flood hydrograph
for n in range(1,nt):
    time[n] = time[n-1] + dt
    if (time[n] <= time_floodpeak):
        Qinlet[n] = (time[n])*(Qpeak-Qinitial)/
    time_floodpeak+Qinitial
        elif (time[n] > time_floodpeak and time[n] <=
    time_floodend):
```

```
            Qinlet[n] = (time_floodend-time[n])/
        (time_floodend-time_floodpeak)\
            *(Qpeak-Qfinal)+Qfinal
        elif (time[n] > time_floodend):
            Qinlet[n] = Qfinal

fig = plt.figure()
ax = fig.add_subplot(1, 1, 1)
ax.plot(time, Qinlet, label='At channel inlet')

# Main program
for n in range(1,nt):
    time[n] = time[n-1] + dt
    Qin = Qinlet[n]
    Anew[0] = (Qin/alpha)**(1/beeta)
#   Computation for Q-s at all point for given time
    for i in range(1, nx):
        Aavg = 0.5*(Aold[i]+Anew[i-1])
        numer = Aold[i]+(alpha*beeta*dtdx*(Aavg**b1)*Anew[i-1])
        denom = 1+(alpha*beeta*dtdx*(Aavg**b1))
        Anew[i] = numer/denom

    Qoutlet[n] = alpha*(Anew[nx-1])**beeta
    Aold = Anew

ax.plot(time, Qoutlet, label='At channel outlet')
ax.set_xlabel('Time (s)')
ax.set_ylabel('Discharge (m3s-1)')
plt.legend()
plt.show()
```

The variables used in this program are similar to those in the previous section for kinematic flow routing in a rectangular channel. However, the inputs new to this program are all related to defining the triangular inflow flood hydrograph, and are described as under:

| Variable | Description | Variable | Description |
|---|---|---|---|
| Qinitial | Discharge value of the flood hydrograph at the beginning (m³/s) | time_ floodpeak | Time to reach the inflow hydrograph peak from the beginning (s) |
| Qpeak | Discharge value of the flood hydrograph at its peak (m³/s) | time_ floodend | Time for the flood hydrograph to end (s) |
| Qfinal | Discharge value of the flood hydrograph at the end of the hydrograph (m³/s) | z | Side slope of the triangular channel |

The above Python program is run with two different discretizations of the length of the channel. When the number of points dividing the channel is 10 (that is, when $nx = 10$, with 9 computational segments), the resulting graphs of the inflow superimposed on the outflow hydrograph is as shown in Figure 4.7(a), while similar graphs plotted for $nx = 100$ is depicted in Figure 4.7(b). It may be noticed that the rise of the flood hydrograph as well as the flood peak in the second case becomes sharper, compared to that of the former. In fact, the peak discharge is also seen to be slightly greater when the number of channel discretization points is increased. Hence, while carrying out numerical simulations, one may check the effect of increasing the computational points; although beyond a certain fineness of the discretized domain, the differences in the resulting flood outflow graphs may not be too discernible. This kind of mesh convergence studies are important for obtaining fairly correct results without making the mesh too fine and consequently increasing the computational time.

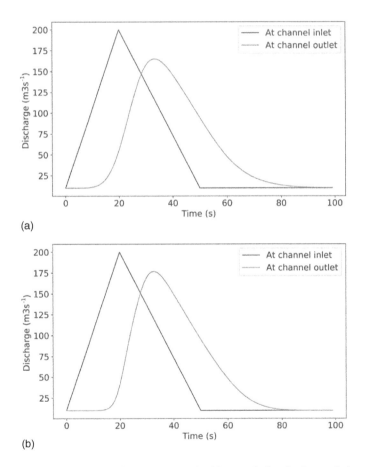

**FIGURE 4.7** Discharges in a triangular channel with same inflow hydrograph, but different channel discretization points. (a) nx = 10 and (b) nx = 100. (Color image available in eBook).

### 4.3.3    Simulation of a Simplified Open-Book Catchment Hydrograph with the Kinematic Wave Approximation

The code for simulating the simplified catchment configuration discussed in Section 4.2.3 (Figure 4.2) is presented below. Here, both the runoffs generated in the overland planes due to a given uniform rainfall and the discharges generated in the central triangular channel that receives the outflows of the catchment planes as lateral inflows are solved using the kinematic wave approximation. The computation algorithms incorporated into the Python program use Equations (4.18) and (4.19), respectively, for the overland runoff and channel flows. The central channel is assumed to be blocked at its upstream end, and is thus considered as its upstream boundary having zero inflow discharge and zero flow depth. The upper edges of the overland planes, similarly, consider the upper boundary condition of zero flow and depth.

```python
# Kinematic wave approximation for open-book catchment
# simulation
import numpy as np
import matplotlib.pylab as plt

z = 2.0; rainfall = 0.001
s0_channel = 0.01; s0_plane = 0.05
mn_channel = 0.02; mn_plane = 0.04
length_channel = 100; length_plane = 50
npoint_channel = 100; npoint_plane = 20
time_rainfall = 250; time_simulation = 500
dt = 1
nt = int(time_simulation/dt)

x_channel = np.linspace(0, length_channel, npoint_channel)
dx_channel = x_channel[1] - x_channel[0]
Aold = np.ones(npoint_channel)*0.001
Anew = np.ones(npoint_channel)*0.001
zz=(z/(1+z*z))**0.3333
alpha_channel = np.sqrt(s0_channel)*zz/mn_channel
beeta_channel = 1.3333
b1 = beeta_channel-1
time = np.zeros(nt)
Qoutlet = np.zeros(nt)

x_plane = np.linspace(0, length_plane, npoint_plane)
dx_plane = x_plane[1] - x_plane[0]
x_channel = np.linspace(0, length_channel, npoint_channel)
dx_channel = x_channel[1] - x_channel[0]

Qold = np.ones(npoint_plane)*0.001
Qnew = np.ones(npoint_plane)*0.001
```

```python
alpha_plane = (mn_plane*1.0**0.6667/np.sqrt(s0_plane))**0.6
beeta_plane = 0.6

dtdx_plane = dt/dx_plane
dtdx_channel = dt/dx_channel
b1 = beeta_plane-1

# Main program
for n in range(1,nt):
    time[n] = time[n-1] + dt
    print("time = ",time[n])
    if (time[n] < time_rainfall):q = rainfall
    else:q = 0.0

    for i in range(1, npoint_plane):
        Qavg = 0.5*(Qold[i]+Qnew[i-1])
        numer = dtdx_plane*Qnew[i-1]+(alpha_plane*beeta_plane*
    Qold[i]*(Qavg**b1))+dt*q
        denom = dtdx_plane+(alpha_plane*beeta_
    plane*(Qavg**b1))
        Qnew[i] = numer/denom

    qlateral = 2*Qnew[npoint_plane-1]
    Qold = Qnew

    for i in range(1, npoint_channel):
        Aavg = 0.5*(Aold[i]+Anew[i-1])
        numer =
Aold[i]+(alpha_channel*beeta_channel*dtdx_channel*(Aavg**b1)*A
  new[i-1])+\
        dt*qlateral
        denom = 1+(alpha_channel*beeta_channel*dtdx_
    channel*(Aavg**b1))
        Anew[i] = numer/denom

    Qoutlet[n] = alpha_channel*(Anew[npoint_channel-1])**
beeta_channel
    Aold = Anew

# print(Qold)
fig = plt.figure()
ax = fig.add_subplot(1, 1, 1)
ax.plot(time, Qoutlet)
ax.set_xlabel('Time (s)')
ax.set_ylabel('Discharge (m3s-1)')

plt.show()
```

The variables used by the above program are described below:

| Variable | Description | Variable | Description |
|---|---|---|---|
| z | Side slopes of the triangular channel (m) | Rainfall | Uniform rainfall intensity (m/s) |
| s0_channel | Longitudinal slope of the channel (-) | mn_ channel | Manning's channel roughness coefficient ($m^{-1/3}s$) |
| s0_plane | Longitudinal slope of the overland planes (-) | mn_plane | Manning's overland roughness coefficient ($m^{-1/3}s$) |
| length_channel | Length of channel (m) | length_ plane | Length of each overland plane (m) |
| npoint_channel | Number of computational points for channel | npoint_ plane | Number of computational points for overland plane |
| timerainfall | Time up to which rainfall occurs (s) | dt | Computational time step (s) |
| timesimulation | Total time up to which simulation is carried up to (s) | | |

The Python code is run for two different rainfall intensities, as was done for the flow taking place through a rectangular channel with lateral inflows (Section 4.3.1). The results (Figure 4.8a and b) again show that the time of saturation of the combined outflow hydrographs vary with increasing rainfall intensities and is achieved relatively soon when intensities are increased.

### 4.3.4 SIMULATION OF A SURGE WAVE IN A TRAPEZOIDAL CHANNEL USING THE ST. VENANT EQUATIONS

The Python program presented in this section simulates the movement of a surge wave in a channel conveying uniform flow (that is, flow at a steady state at normal depth of flow) when a downstream control gate of the channel is suddenly closed. Equations (4.20) and (4.21), mathematically describing the solution of the St. Venant flow equations using the Preissmann implicit finite difference scheme, are implemented in the code in their derivative forms, as required for solving the system of non-linear equations by the method of N-R. The variables used are $y$ for depth, and $v$ for velocity. The initial uniform depth of flow is computed manually from the given discharge and defined as the initial condition for all the nodes to start the computations. The boundary conditions, incorporated within the time simulation portion of the code, are defined as follows:

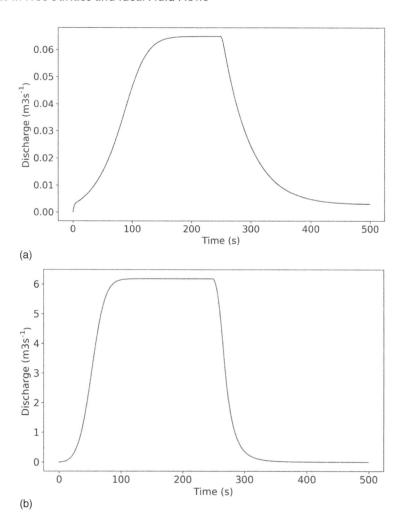

(a)

(b)

**FIGURE 4.8**  Outflow discharges from the "open-book" catchment for: (a) rainfall intensity of $0.0001$ ms$^{-1}$ and (b) rainfall intensity of $0.001$ ms$^{-1}$. (Color image available in eBook).

1. Upstream boundary: Constant flow depth, expressed as N-R equivalent
2. Downstream boundary: Zero flow velocity, expressed as N-R equivalent whereby the right hand side of the equation is specified as the negative of the velocity at this node.

```
# Unsteady, free-surface open channel flow computation by
# Preissmann scheme
import numpy as np
import matplotlib.pylab as plt
from numpy import array
g = 9.81
nx = 201
```

```python
time_simulation = 200
length = 1000
b0 = 10.0
s = 2.0
s0 = 0.0001
mn = 0.01
Q = 100
y0 = 3.426
theta = 0.8
errorallow = 0.0001
maxiter = 10
dt = 1.0

def area(depth):
    area = (b0+depth*s)*depth
    return area

def wetperi(depth):
    wetperi = b0+2*depth*np.sqrt(1+s*s)
    return wetperi

def hyrad(depth):
    hyrad = (b0+depth*s)*depth/(b0+2*depth*np.sqrt(1+s*s))
    return hyrad

def topwidth(depth):
    topwidth = (b0+2*depth*s)
    return topwidth

def centroid(depth):
    centroid = depth*depth*(b0/2+depth*s/3)
    return centroid

def dcendy(depth):
    dcendy = depth*(b0+depth*s)
    return dcendy

nt = int(time_simulation/dt)

xx = np.linspace(0, length, nx)
dx = xx[1]-xx[0]
dtdx = dt/dx
v0 = Q/area(y0)
y = np.ones(nx)*y0
v = np.ones(nx)*v0
ca = np.zeros(nx)
cb = np.zeros(nx)
A = np.zeros((2*nx,2*nx))
b = np.zeros(2*nx)
delx = np.zeros(2*nx)
time = np.zeros(nt)
```

```
for n in range (1,nt):

    time[n] = time[n-1] + dt

    for i in range(0,nx-1):
        area1 = area(y[i])
        area2 = area(y[i+1])
        ca[i] = 2*dtdx*(1-theta)*(area2*v[i+1]-area1*v[i])-
    (area1+area2)
        sf1 = abs(v[i])*v[i]*mn**2/hyrad(y[i])**1.3333
        sf2 = abs(v[i+1])*v[i+1]*mn**2/hyrad(y[i+1])**1.3333
        aa = (-1.0)*dt*(1-theta)*(g*area2*(s0-
    sf2)+g*area1*(s0-sf1))
        bb = (-1.0)*(v[i]*area1+v[i+1]*area2)
        cc = 2*dtdx*(1-theta)*(v[i+1]**2*area2+g*centroid(
    y[i+1])\
                        -v[i]**2*area1-g*centroid(y[i]))
        cb[i] = aa+bb+cc

    iter = 0
    max_error = 1.0
    while (max_error > errorallow):
        iter = iter+1
        if(iter > maxiter): break
        # Boundary condition at upstream node
        A[0,0] = 1.0
        b[0] = (-1.0)*(y[0]-y0)
        # Boundary condition at downstream node
        A[2*nx-1,2*nx-1] = 1.0
        b[2*nx-1] = (-1.0)*(v[nx-1])
        # Forming equations for the interior nodes
        for i in range(0,nx-1):
            n = 2*i
            area1 = area(y[i])
            area2 = area(y[i+1])
            b[n+1] = (-1.0)*(area1+area2+2*dtdx*theta\
              *(v[i+1]*area2-v[i]*area1)+ca[i])
            sf1 = abs(v[i])*v[i]*mn**2/hyrad(y[i])**1.3333
            sf2 = abs(v[i+1])*v[i+1]*mn**2/
    hyrad(y[i+1])**1.3333
            aa = 2*dtdx*theta*(v[i+1]**2*area2+g*centroid
    (y[i+1])\
            -(v[i]**2*area1+g*centroid(y[i])))
            bb = (-1.0)*theta*dt*g*((s0-sf2)*area2+(s0-
    sf1)*area1)
            b[n+2] = (-1.0)*(v[i]*area1+v[i+1]*area2+aa+bb+c
    b[i])
            dady1 = topwidth(y[i])
            dady2 = topwidth(y[i+1])
            A[n+1,n] = dady1*(1-2*dtdx*theta*v[i])
            A[n+1,n+1] = (-1.0)*2*dtdx*theta*area1
```

```
            A[n+1,n+2] = dady2*(1+2*dtdx*theta*v[i+1])
            A[n+1,n+3] = 2*dtdx*theta*area2
            dcdy1 = dcendy(y[i])
            dcdy2 = dcendy(y[i+1])
            dsdv1 = (v[i]+abs(v[i]))*mn**2/
    (hyrad(y[i]))**1.3333
            dsdv2 = (v[i+1]+abs(v[i+1]))*mn**2/
    (hyrad(y[i+1]))**1.3333
            perimeter1 = wetperi(y[i])
            aa1 = 2*np.sqrt(1+s**2)*area1-dady1*perimeter1
            bb1 = (hyrad(y[i]))**0.3333*area1**2
            dsdy1 = 1.3333*v[i]*abs(v[i])*mn**2*aa1/bb1
            perimeter2 = wetperi(y[i+1])
            aa2 = 2*np.sqrt(1+s**2)*area2-dady2*perimeter2
            bb2 = (hyrad(y[i+1]))**0.3333*area2**2
            dsdy2 = 1.3333*v[i+1]*abs(v[i+1])*mn**2*aa2/bb2
            cc1 = (-1.0)*2*dtdx*theta*(v[i]**2*dady1+g*dcdy1)
            dd1 = (-1.0)*g*dt*theta*(s0-sf1)*dady1
            cc2 = 2*dtdx*theta*(v[i+1]**2*dady2+g*dcdy2)
            dd2 = (-1.0)*g*dt*theta*(s0-sf2)*dady2
            A[n+2,n] = v[i]*dady1+cc1+dd1+g*dt*theta*area1*d
    sdy1
            A[n+2,n+1] = area1-2*dtdx*theta*2*v[i]*area1+g*dt*
    theta*area1*dsdv1
            A[n+2,n+2] = v[i+1]*dady2+cc2+dd2+g*dt*theta*area
    2*dsdy2
            A[n+2,n+3] = area2+2*dtdx*theta*2*v[i+1]*area2+g*d
    t*theta*area2*dsdv2

        delx = np.linalg.solve(A,b)
        max_error = np.max(abs(delx))
        for i in range(0,nx):
            y[i] += delx[2*i]
            v[i] += delx[2*i+1]

fig = plt.figure()
ax = fig.add_subplot(1, 1, 1)
ax.plot(xx/1000, y)
ax.set_xlabel('Distance from upstream end (km)')
ax.set_ylabel('Water surface (m)')
plt.show()
```

The variables used in the program are as follows:

| Variable | Description | Variable | Description |
|---|---|---|---|
| g | Acceleration due to gravity $(ms^{-2})$ | b0 | Bed width of the trapezoidal channel (m) |
| length | Length of channel (m) | s | Side slope of the trapezoidal channel (-) |
| nx | Number of computational points for the channel | s0 | Longitudinal slope of the channel (-) |

*(Continued)*

| Variable | Description | Variable | Description |
|----------|-------------|----------|-------------|
| Q | Discharge in the channel (m³s⁻¹) | y0 | Normal depth in channel corresponding to Q (m) |
| theta | Time weighting coefficient in Preissmann scheme | errorallow | Error tolerance in N-R computations (same for $\Delta y$ and $\Delta v$; m and ms⁻¹) |
| dt | Computational time step (s) | maxiter | Maximum iterations allowed in N-R computations |

The Python code computes the simulations through time and plots the water surface profile along the length of the channel at specified times. Figures 4.9(a) and (b) show such water surface profiles of the travelling surge wave at two different instances of time.

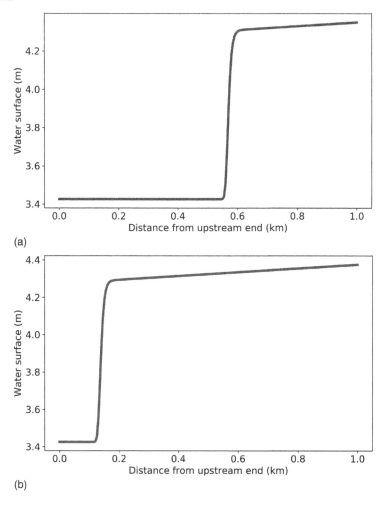

(a)

(b)

**FIGURE 4.9**  Water surface profiles plotted against the length of the 1 km long trapezoidal channel generated by a positive surge wave moving upstream at times: (a) 100 s from beginning and (b) 200 s from beginning. (Color image available in eBook).

The progress of the surge front may be observed from Figure 4.9, in which the front is seen to move forward (opposite to the flow direction). On reaching the left-most boundary, where a constant water level is specified the wave starts travelling downstream, as a negative surge front. Although still images have been shown in the figures, suitable Python animation routines may be invoked to dynamically visualize the movement of the surge wave over the complete simulation time.

### 4.3.5  SIMULATION OF STREAMLINES IN AN IDEAL FLUID FLOW

The simulations depicted in the previous sections were governed by PDEs in terms of the $x$- and $t$-variables, which are associated with one-dimensional unsteady flow. In this example, the steady-state streamlines for a two-dimensional ideal irrotational fluid flow (Section 4.2.5) is obtained and plotted with the help of the equivalent difference equation given by Equation (4.22), and appropriate boundary conditions, such as that given in Equation (4.23). The following program in Python illustrates the method for the flow taking place through a contracted section of a conduit, discussed in Section 4.2.5.

```python
# Streamlines in irrotational flow
import numpy as np
import matplotlib.pyplot as plt

nx = 25
ny = 50
n = nx*ny
stepheight = 10
jstep = 30

A = np.zeros((n,n))
c = np.zeros(n)

for i in range(0,nx):
    A[i,i] = 1.0
    c[i] = (nx-i-1)*1.0

for j in range(1,ny):
    A[nx*j,nx*j] = 1.0
    c[nx*j] = nx-1
    if(j<jstep):step = 0
    else: step = int(stepheight)
    A[nx*(j+1)-1-step,nx*(j+1)- 1-step] = 1.0
    c[nx*(j+1)-1-step] = step

step = int(stepheight)
for j in range(jstep,ny):
    for i in range(nx-step-1,nx):
        nn = nx*j+i
        A[nn,nn] = 1.0
        c[nn] = 0.0
```

```
for i in range(1,nx-step-1):
    A[nx*(ny-1)+i,nx*(ny-1)+i] = 1.0
    A[nx*(ny-1)+i,nx*(ny-2)+i] = -1.0
    c[nx*(ny-1)+i] = 0.0

for j in range(1,ny-1):
    if(j<jstep):
        for i in range(1,nx-1):
            nn = nx*j+i
            A[nn,nn] = -4
            A[nn,nn-1] = 1
            A[nn,nn+1] = 1
            A[nn,nn-nx] = 1
            A[nn,nn+nx] = 1
    else:
        for i in range(1,nx-step-1):
            nn = nx*j+i
            A[nn,nn] = -4
            A[nn,nn-1] = 1
            A[nn,nn+1] = 1
            A[nn,nn-nx] = 1
            A[nn,nn+nx] = 1

x = np.linalg.solve(A,c)

Z = np.zeros((nx,ny))
for j in range(0,ny):
    for i in range(0,nx):
        Z[nx-i-1,j] = x[j*nx+i]

xlist = np.linspace(0, nx-1, nx)
ylist = np.linspace(0, ny-1, ny)
X, Y = np.meshgrid(ylist, xlist)

fig, ax = plt.subplots(figsize=(8,4))
cp = ax.contour(X, Y, Z, 20, cmap='coolwarm')
ax.set_title('Streamlines', fontsize=15)
ax.set_xlabel('x (m)', fontsize=15)
ax.set_ylabel('y (m)', fontsize=15)
plt.show()

nx = 25
ny = 50
stepheight = 10
jstep = 30
```

The variables used in the program are given below, and are related to the problem definition given in Section 4.2.5:

| Variable | Description | Variable | Description |
|----------|-------------|----------|-------------|
| nx | Discretization points in the x-direction (i-index) | stepheight | Height of step in i-index units |
| ny | Discretization points in the y-direction (j-index) | jstep | Location of step in j-index units from j=0 |

The program, when run, plots the streamlines in the flow region as shown in Figure 4.3.

### 4.3.6 TWO-DIMENSIONAL DEPTH-AVERAGED FLOWS IN A SHALLOW BASIN

The Python program for finding the flow velocities in a shallow rectangular water basin is given below. The basin, schematically depicted in Figure 4.10a, is assumed as having a horizontal bed, with length 1000 m and width 500 m divided into grids of size 10 m in either direction. Thus there are 101 grid points along the length of the basin, including the boundaries at both end and 51 such points along the width. It is assumed that a constant inflow enters the basin through 5 boundary nodes, as shown in Figure 4.10b. The inflow discharge is assumed as 1 $m^3s^{-1}$ for each of the 5 inlet nodes. However, the value is converted to equivalent velocity while implementation it in the code. The excess water is assumed to flow out through a weir outlet on the opposite edge of the water body, spread over 10 grid points. The crest level of the weir is assumed to be 2.0 m above the bed of the basin. The standard weir flow equation for discharge $Q$ ($Q = C_dLH^{3/2}$) is used while applying the outflow-end boundary condition in the form of a relation between the depth of flow above the weir crest and the velocity at the neighbouring interior point. In the discharge formula, $C_d$ is the coefficient of discharge (assumed to be 2.2) and $H$ is the depth of water above the weir crest level. On simplifying, $H$ may approximately be expressed in terms of the velocity $v$, as $H = 2.0v^2$. The gross flow depth is then given as the weir height (2 m) + $H$ (in m). The Manning's flow friction coefficient is assumed to be 0.1 $m^{-1/3}s$ for all the grid points.

Note that in the program, the following notations are used:

| Notation | Description | Notation | Description |
|----------|-------------|----------|-------------|
| U | Velocity in the x-direction (i-index) | uh | The variable u*h, stored as an array |
| V | Velocity in the y-direction (j-index) | vh | The variable v*h, stored as an array |
| H | Depth of flow | | |

The program presented below in Python is inspired by a similar code developed by Dr. Paul Connolly (Connolly, 2017) for simulating gravity waves in the atmosphere. Although the algorithm for solving the equations are the same, that is the Lax–Wendroff technique as explained in Section 4.2.6, the main difference between Dr. Connolly's code and the one given here lies at the implementation of the

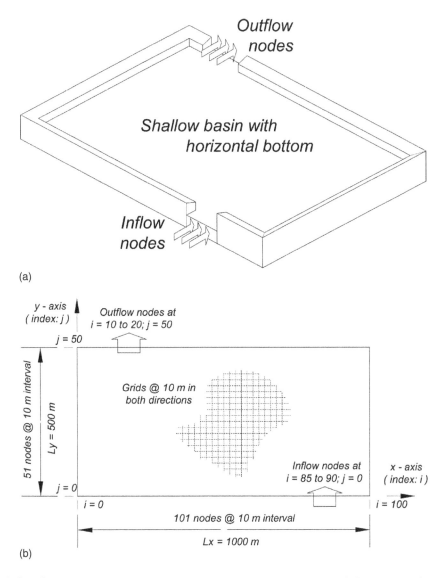

**FIGURE 4.10** Rectangular basin for computation of circulation due to inflows and outflows. (a) schematic view of the basin; (b) computational grid showing locations of inflow and outflow points.

boundary conditions and the inclusion of the frictional force on the flow to make it adaptable for free surface flows. The program may further be modified to implement other numerical algorithms for its solution.

```
# Simulation of shallow tank with inflow through 5 nodes and
# outflow through 10
#
```

```python
import numpy as np
import matplotlib.pyplot as plt

fou = open("u.csv","w")
fov = open("v.csv","w")

g = 9.81
nx = 101
ny = 51
a = 10
dt = 0.01
time_simulation = 250
Qin = 1.0
mn = 0.10

nt = int(np.fix(time_simulation/dt)+1)

x=np.mgrid[0:nx]*a
y=np.mgrid[0:ny]*a
[Y,X] = np.meshgrid(y,x)

z = np.zeros((nx, ny))
height = 2.0*np.ones((nx, ny))
u=np.zeros((nx, ny))
v=np.zeros((nx, ny))

h = height - z
h_rhs = np.zeros((nx-2, ny-2))
for i in range(85,90):
    h_rhs[i,0] = Qin/(a*a)

# Subroutine for Lax-Wendroff algorithm for interior grid
# points

def lax_wendroff(dd, dt, g, u, v, h, usource, vsource,
hsource):

    uh = u*h
    vh = v*h

    h_mid_xt = 0.5*(h[1:,:]+h[0:-1,:])-(0.5*dt/
dd)*(uh[1:,:]-uh[0:-1,:])
    h_mid_yt = 0.5*(h[:,1:]+h[:,0:-1])-(0.5*dt/
dd)*(vh[:,1:]-vh[:,0:-1])

    Ux = uh*u+0.5*g*h**2
    Uy = uh*v
    uh_mid_xt = 0.5*(uh[1:,:]+uh[0:-1,:])-(0.5*dt/
dd)*(Ux[1:,:]-Ux[0:-1,:])
    uh_mid_yt = 0.5*(uh[:,1:]+uh[:,0:-1])-(0.5*dt/
dd)*(Uy[:,1:]-Uy[:,0:-1])
```

```
    Vx = Uy
    Vy = vh*v+0.5*g*h**2.
       vh_mid_xt = 0.5*(vh[1:,:]+vh[0:-1,:])-(0.5*dt/
    dd)*(Vx[1:,:]-Vx[0:-1,:])
       vh_mid_yt = 0.5*(vh[:,1:]+vh[:,0:-1])-(0.5*dt/
    dd)*(Vy[:,1:]-Vy[:,0:-1])

    h_new = h[1:-1,1:-1] \
       - (dt/dd)*(uh_mid_xt[1:,1:-1]-uh_mid_xt[0:-1,1:-1]) \
       - (dt/dd)*(vh_mid_yt[1:-1,1:]-vh_mid_yt[1:-1,0:-1]) \
       + dt*hsource

       Ux_mid_xt = uh_mid_xt*uh_mid_xt/h_mid_xt +
    0.5*g*h_mid_xt**2.
    Uy_mid_yt = uh_mid_yt*vh_mid_yt/h_mid_yt
    uh_new = uh[1:-1,1:-1] \
       - (dt/dd)*(Ux_mid_xt[1:,1:-1]-Ux_mid_xt[0:-1,1:-1]) \
       - (dt/dd)*(Uy_mid_yt[1:-1,1:]-Uy_mid_yt[1:-1,0:-1]) \
       + dt*usource*0.5*(h[1:-1,1:-1]+h_new)

    Vx_mid_xt = uh_mid_xt*vh_mid_xt/h_mid_xt
       Vy_mid_yt = vh_mid_yt*vh_mid_yt/h_mid_yt +
    0.5*g*h_mid_yt**2.
    vh_new = vh[1:-1,1:-1] \
       - (dt/dd)*(Vx_mid_xt[1:,1:-1]-Vx_mid_xt[0:-1,1:-1]) \
       - (dt/dd)*(Vy_mid_yt[1:-1,1:]-Vy_mid_yt[1:-1,0:-1]) \
       + dt*vsource*0.5*(h[1:-1,1:-1]+h_new)
    u_new = uh_new/h_new
    v_new = vh_new/h_new
    # Returning the updated variable values
    return (u_new, v_new, h_new)

# Main program
for n in range(0,nt):
    sqrtu2v2 = np.sqrt(u*u+v*v)
       u_rhs = (g*mn**2*(u[1:-1,1:-1])*sqrtu2v2[1:-1,1:-1])/
    (h[1:-1,1:-1])**1.33
       v_rhs = (g*mn**2*(v[1:-1,1:-1])*sqrtu2v2[1:-1,1:-1])/
    (h[1:-1,1:-1])**1.33

       # Run and obtain updated variable values
       (unew, vnew, hnew) = lax_wendroff(a, dt, g, u, v, h,
    u_rhs, v_rhs, h_rhs)

     u[1:-1,1:-1] = unew;
     v[1:-1,1:-1] = vnew;

     u[[0,-1],:]=0.;
     v[0,1:-1]=vnew[0,:]
     v[-1,1:-1]=vnew[-1,:]
```

```
        v[:,[0,-1]]=0.;
        u[1:-1,0]=unew[:,0]
        u[1:-1,-1]=unew[:,-1]
        h[1:-1,1:-1] = hnew;
        h[0,1:-1]=hnew[0,1]
        h[-1,1:-1]=hnew[-1,:]
        h[1:-1,0]=hnew[:,0]
        h[1:-1,-1]=hnew[:,-1]
        h[0,0]=hnew[0,0]
        h[1,-1]=hnew[1,-1]
        h[-1,0]=hnew[-1,0]
        h[-1,-1]=hnew[-1,-1]

        for i in range(10,20):
            h[i,-1] = 2.0+0.2*(v[i,-2])**2

fig, ax = plt.subplots(figsize=(10,5))
ax.quiver(X,Y,u,v)
plt.show()

np.savetxt('u.csv', u, '%10.5f', delimiter=',')
np.savetxt('v.csv', v, '%10.5f', delimiter=',')

fou.close()
fov.close()
```

The variables used in the program are as given below:

| Variable | Description | Variable | Description |
|---|---|---|---|
| g | Acceleration due to gravity $(ms^{-2})$ | dt | Computational time step (s) |
| mn | Manning's roughness coefficient $(m^{-1/3}s)$ | time_simulation | Time till which simulation is done (s) |
| a | Size of computational cell in x- and y-directions (m) | nx | Discretization points in the x-direction (i-index) |
| Qin | Inflow discharge through each inlet point $(m^3s^{-1})$ | ny | Discretization points in the y-direction (j-index) |

The following points explain the key features of the code:

1. At the beginning of the code, two .csv files are opened, one each to store the velocity values *u* and *v* that are computed at the end of the simulations.
2. The number of grid points (nx and ny) in the *x*- and *y*-directions are next defined that serve as the computational grid points and also help in plotting the velocity vectors at the end of the simulation.

3. Next, the following variables z, height, u, v, and h are initiated, all having the shape (nx, ny). Here, $z$, the bed elevation defined above an assumed datum, is initiated to zero as the bottom or bed of the shallow basin is assumed to be flat. If a spatially varying bathymetry is to be incorporated, then it has to be accordingly specified through this rectangular array. Additionally, the gravity component term due to a sloping bottom also has to be included on the right hand side source terms of the $x$- and $y$-momentum equations of the code, which have been presently assumed to be zero in the code. The variable height defines the water surface above the assumed datum.

4. An array h_rhs is defined which stores the source terms at each grid point. Note that this variable is defined at all the interior nodes for which the Lax–Wendroff calculations are run. This array is set to zero at all computable grid nodes, except at the inflow boundary nodes where the inflow discharge source terms (Qin/(a*a)) are set to the actual values.

5. The Lax–Wendroff calculations are defined within a Python function, similar to a subroutine of FORTRAN, and accepts the input variables dd, dt, g, u, v, h, usource, vsource, and hsource, which stand for the following terms, respectively: grid spacing; time interval for computation; acceleration due to gravity; the arrays of $x$-velocity, $y$-velocity, depth of flow (as arrays of size nx by ny); and the $x$-momentum, $y$-momentum, and continuity equation source terms (size nx−2 by ny−2) of Equations (4.24). The subroutine returns the variables u_new, v_new, and h_new (each of size nx−2 by ny−2), which are the updated velocity and flow depth values at the computable grid nodes.

6. The main program is defined at the end of the code and is essentially a loop over the time steps, starting from the initial conditions. At each time level, the source terms for the momentum equations (u_rhs and v_rhs) are first computed. Note that those for the continuity equation, defined outside the time-loop, remain unchanged over the time iterations. The u_rhs and

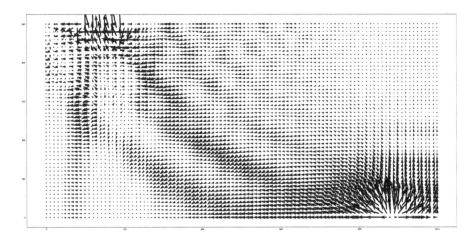

**FIGURE 4.11** Flow vectors (quiver-plot) of the shallow basin flow simulation problem, solved by the Lax–Wendroff method, at the end of 250 s of simulation.

v_rhs variables are both of size nx-2 by ny-2. The Lax–Wendroff function is then invoked and the updated velocity and flow depth variables returned by the function are accepted by the main program and used to update the global variables. Note that the outlet boundary condition is also implemented at this stage as the depth of flow at the outlet nodes are set according to the weir outlet equations assumed for this problem.

7. Once complete, the main program is exited and a quiver-plot showing the velocity vectors, as shown in Figure 4.11, is generated as output. Also, the final velocities computed in the $x$- and $y$-directions are saved in the two files u.csv and v.csv, which are used for estimating the spread of a contaminant within the shallow water body in Chapter 6.

## REFERENCES

Chaudhry, M. H. (2008). *Open Channel Flow*. Springer, 2nd edition.

Chow, V., Maidment, D. and Mays, L. (1988) *Applied Hydrology*. McGraw-Hill Book Company, New York.

Connolly.(2017). *Shallow Water Equations in MATLAB / Python*. Viewed on April 17 2021. https://personalpages.manchester.ac.uk/staff/paul.connolly/teaching/practicals/shallow_water_equations.html

Lax, P. D. and Wendroff, B. (1960). "Systems of conservation laws". *Commun. Pure Appl. Math.*, 13, 217–237.

Wooding, R. A. (1965). "A Hydraulic Model for the Catchment-Stream Problem", *J. Hydrol.*, 3, 254–267.

Vreugdenhil, C. B. (1989). *Computational Hydraulics: An Introduction*. Springer Verlag.

# 5 Partial Differential Equations in Subsurface Flows

Subsurface flows, or the movement of water occurring below the ground surface, cover a wide range of phenomena and are mostly three-dimensional in nature. However, for simplicity, some subsurface flows may conveniently be approximated in the two-dimensional plane, because of the predominance of the flow velocity directions in such a plane which maybe oriented either vertically or horizontally. The former is used generally for handling seepage problems, like the flow below a weir across a river located on a pervious foundation and holding back a high depth of water and having a lower water depth on the downstream. Seepage flow also occurs through earthen embankments or earth-retaining piles or walls having different water levels on either side. Subsurface flows through shallow aquifers, also known as groundwater flows, may be approximated as taking place in the horizontal plane, with the velocities averaged over its depth at any point. The flow here is driven by the difference in water levels across its boundaries but may also be influenced by the inflows infiltrating into the aquifer from above by rainwater recharge or because of outflows due to withdrawal of water by pumping from wells. Numerical techniques using the method of finite differences for solving the problems in seepage and groundwater flows are presented in this chapter, following a description of the governing equations of the physical processes. Corresponding Python programs implementing these solution techniques are given in the final section of this chapter.

## 5.1 GOVERNING EQUATIONS OF SUBSURFACE FLOWS

The Darcy's law describes the fundamental equation for computing flows through a porous medium which, in the three perpendicular coordinate directions $x$, $y$, and $z$, may be written in the following manner (Wang and Anderson, 1982):

$$q_x = -K\frac{\partial h}{\partial x}; q_y = -K\frac{\partial h}{\partial y}; \text{ and } q_z = -K\frac{\partial h}{\partial z} \qquad (5.1)$$

In the above equations, the notations used are: $q_x$, $q_y$, and $q_z$, the volume flow rate per unit area in the three directions; and $K$, the hydraulic conductivity of the medium, assumed to be isotropic in this book for simplicity. The space derivatives in Equation (5.1) are the partial derivatives of the piezometric head or the depth of saturated water column within the soil above a given point, $h$, with respect to the three

DOI: 10.1201/9780429288579-5

coordinate axes. The expressions in Equations (5.1), which relate the velocities in the three coordinate directions to the respective gradients of the pressure heads, may be substituted in the corresponding equation of continuity for steady state to obtain the following:

$$\frac{\partial q_x}{\partial x} + \frac{\partial q_y}{\partial y} + \frac{\partial q_z}{\partial z} = 0 \tag{5.2}$$

Since $K$, the hydraulic conductivity, is assumed to be independent on the coordinate axes, the following equation results are obtained after simplification (Equation 5.3). This is also called the Laplace equation, and is applied as the governing equation of groundwater flow taking place through an isotropic, homogenous soil mass under steady-state conditions.

$$\frac{\partial^2 h}{\partial x^2} + \frac{\partial^2 h}{\partial y^2} + \frac{\partial^2 h}{\partial z^2} = 0 \tag{5.3}$$

A few representative problems that may be solved using the above equations are shown in Figure 5.1.

### 5.1.1 Governing Equations of Flow in an Unconfined Aquifer

In this case, the Dupuit assumptions (Wang and Anderson, 1982) are used, which approximates the three-dimensional nature of the flow to an essentially two-dimensional flow in the horizontal plane. The hydraulic gradient, in this case, may be estimated from the slope of the free surface. By considering a differential area in plan, it may be shown that if the recharge or inflow to it is given as $R$, in the units of volumetric flow rate per unit time per unit area, the governing equation (Equation 5.3) may be expressed as:

$$\frac{K}{2}\left( \frac{\partial^2 \left( h^2 \right)}{\partial x^2} + \frac{\partial^2 \left( h^2 \right)}{\partial y^2} \right) = -R \tag{5.4}$$

Instead of recharge, if water is withdrawn or pumped out from the elemental control area, the value of $R$ may be taken as negative. Equation (5.4), which is valid for a steady-state condition, may be further modified for an unsteady condition by incorporating the storage term, as under:

$$\frac{K}{2}\left( \frac{\partial^2 \left( h^2 \right)}{\partial x^2} + \frac{\partial^2 \left( h^2 \right)}{\partial y^2} \right) = S\frac{\partial h}{\partial t} - R \tag{5.5}$$

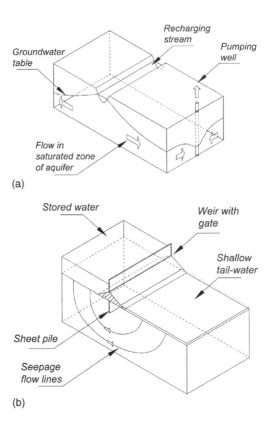

FIGURE 5.1   (a) An aquifer being recharged from a stream and discharged by a pumping well; (b) seepage flow caused by water held back by a weir with a raised gate.

In the above equation, $S$ is the storage coefficient or the specific yield of the unconfined aquifer, which is equivalently the volume of water that is released from storage per unit area (in plan) per unit decline in the pressure head, $h$.

### 5.1.2   GOVERNING EQUATIONS OF FLOW IN A CONFINED AQUIFER

For a horizontal aquifer the volumetric rates flow per unit time per unit width of the aquifer in the two coordinate directions $x$ and $y$, $q_x$ and $q_y$, may be expressed as follows:

$$q_x = -KD\frac{\partial h}{\partial x} \tag{5.6a}$$

$$q_y = -KD\frac{\partial h}{\partial y} \tag{5.6b}$$

In the above equation, $h$ is the piezometric head within the confined aquifer at a location and $D$ is the thickness of the aquifer medium. When the expressions in Equations (5.6a and 5.6b) are substituted in the continuity equation with the product $KD$ defined as the transmissivity, $T$, of the aquifer, and assuming a recharge $R$ into the system, the following form of the governing equation for a confined aquifer is obtained for the unsteady condition:

$$\frac{\partial^2 h}{\partial x^2} + \frac{\partial^2 h}{\partial y^2} = \frac{S}{T}\frac{\partial h}{\partial t} - \frac{R}{T} \tag{5.7}$$

### 5.1.3 GOVERNING EQUATION OF STEADY-STATE SEEPAGE IN THE VERTICAL PLANE

Seepage in the vertical plane is an important consideration in quite a few engineering problems such as flow through an embankment dam, or flow below a weir. In each case, the governing equation in the two-dimensional form of Equation (5.3) is given as

$$\frac{\partial^2 h}{\partial x^2} + \frac{\partial^2 h}{\partial z^2} = 0 \tag{5.8}$$

Here, a steady-state situation is assumed and the horizontal and vertical coordinate directions are designated as $x$ and $z$, respectively. However, the solutions would be dependent on the boundary conditions specified for the problem. In most cases, these may either be in the form of a specific head, $h$, or specified as no-flow boundary which translates mathematically as $\frac{\partial h}{\partial n} = 0$, where $n$ is the direction normal to the edge of the boundary. The first of the two is known as the Dirchlet-type, and the other as the Neumann-type, of boundary conditions.

## 5.2 NUMERICAL METHODS FOR SOLVING THE GROUNDWATER AND SEEPAGE FLOW EQUATIONS

The equations on groundwater and seepage flows are second-order partial differential equations, for which the finite difference schemes are used to solve them approximately. Both explicit and implicit schemes may be used, as shown in Wang and Anderson (1982), but for the examples discussed here we implement the latter in all cases. Further, these examples are limited to one- and two-dimensional spaces. For the former, the one-dimensional domain is divided equally into a number of segments connected at computational nodes for implementation of the finite difference method. For the two-dimensional problems – spread horizontally for groundwater flows or vertically for seepage flows – the region is discretized into suitable rectangular computational grids distributed in either direction.

### 5.2.1 SOLVING THE UNSTEADY ONE-DIMENSIONAL GROUNDWATER FLOW IN AN UNCONFINED AQUIFER

The problem discussed in Section 3.1.6 of one-dimensional steady-state groundwater flow profile between two ditches, is modified here to a transient or unsteady case by assuming the water level in one of the ditches to drop from an initial steady-state value and causing the groundwater profile to gradually modify itself to the changed condition with time. The corresponding equation, which now becomes a partial differential equation in variables $x$ and $t$, is non-linear due to the presence of the $h^2$ term in the differential and is written as under:

$$\frac{K}{2} \frac{\partial^2 (h^2)}{\partial x^2} = S \frac{\partial h}{\partial t} \mp R \tag{5.9}$$

The above equation, in which a positive value of $R$ implies recharge, may also be written as:

$$\frac{K}{2} \frac{\partial^2 (h^2)}{\partial x^2} = \frac{S}{2h} \frac{\partial (h^2)}{\partial t} \mp R \tag{5.10}$$

Here we use a one-dimensional computational domain for which we discretize the above equation by the following differencing procedure:

$$\left[ \frac{(h^2)_{i-1}^{n+1} - 2(h^2)_i^{n+1} + (h^2)_{i+1}^{n+1}}{(\Delta x)^2} \right] = \left( \frac{1}{K} \right) \left( \frac{S}{h_i} \right) \frac{(h^2)_i^{n+1} - (h^2)_i^{n}}{\Delta t} \mp 2R \left( \frac{1}{K} \right) \tag{5.11}$$

In Equation (5.11), the subscripts denote the computational node around which the difference equations are developed while the superscripts, $n$ and $n+1$, denote the known or current and the next or unknown time levels, respectively. Also, $\Delta x$ and $\Delta t$ are the space and time steps, respectively. A simplification of the first term on the right may be made in the following way, which will not drastically affect the final solution:

$$\left( \frac{1}{K} \right) \left( \frac{S}{h_i} \right) \frac{(h^2)_i^{n+1} - (h^2)_i^{n}}{\Delta t} = \left( \frac{S}{K(\Delta t)} \right) \left[ \frac{(h^2)_i^{n+1}}{(h)_i^{n+1}} - \frac{(h^2)_i^{n}}{(h)_i^{n}} \right] \tag{5.12}$$

Thus, Equation (5.12) may be written in the form of a non-linear function of the depths $h$ at the nodes $i$-1, $i$, and $i+1$, as under:

$$f\left( h_{i-1}^{n+1}, h_i^{n+1}, h_{i+1}^{n+1} \right) = (h^2)_{i-1}^{n+1} - 2(h^2)_i^{n+1} + (h^2)_{i+1}^{n+1} - \left( \frac{(\Delta x)^2}{K} \right) \left( \frac{S}{\Delta t} \right) (h)_i^{n+1}$$
$$- \left[ \left( \frac{(\Delta x)^2}{K} \right) \left( \frac{S}{\Delta t} \right) (h^2)_i^{n} \pm 2R \left( \frac{(\Delta x)^2}{K} \right) \right] \tag{5.13}$$

The terms containing the depth $h$ at node $i$ and time level $n$ is considered known from the previous time condition. The above equation may now be solved using the Newton–Raphson (N-R) method for a system of non-linear equations.

## 5.2.2   SOLVING THE UNSTEADY TWO-DIMENSIONAL GROUNDWATER FLOW IN AN UNCONFINED AQUIFER

The governing equation for two-dimensional unsteady flow in an unconfined aquifer, Equation (5.5), is simplified in this case as:

$$S\frac{\partial h}{\partial t} = K\left[\frac{\partial}{\partial x}\left(d\frac{\partial h}{\partial x}\right) + \frac{\partial}{\partial y}\left(d\frac{\partial h}{\partial y}\right)\right] \pm R \qquad (5.14)$$

In Equation (5.14), the terms are similar to those used previously in this section. The newly introduced term, $d$, stands for the saturated depth through which the flow is taking place and is the difference between the groundwater table level and the bed elevation of the aquifer at that point. The generalized groundwater flow problem over an uneven topography, may be shown as in Figure 5.2a, wherein a two-dimensional computational grid is considered over the surface in the horizontal $x$-$y$ plane, with the grid spacing in the respective directions designated as $\Delta x$ and $\Delta y$. The water table is assumed to have an average level within each cell, having an elevation $h$, measured from a horizontal datum. Similarly, the bottom impervious surface of the aquifer is assumed to be horizontal within each cell, although its elevation above a common datum may vary from cell to cell. This value is designated by $z$, also measured from the same datum as that of $h$.

A typical cluster of 5 cells, with the central cell having the index $(i,j)$ is shown in Figure 5.2(b). The neighbouring cells are numbered according to the scheme shown in Figure 5.2(a). The grid cell size in each direction is assumed to be the same (that is, $\Delta x = \Delta y$) and equal to $a$. The central computational cell $(i,j)$ and the inflows into it from the four neighbouring cells are shown in Figure 5.2(c). Following McWhorter and Sunada (1977), we may denote the flow taking place from a typical neighbouring cell, say cell $(i\text{-}1,j)$, into the central cell $(i,j)$ by the notation $Q_{i-1/2,j}$ and write the equation as below:

$$Q_{i-1/2,j} = aKd_{i-1/2}^{t}\left[\frac{\left(h_{i-1,j} - h_{i,j}\right)^{t+\Delta t}}{a}\right] \qquad (5.15a)$$

In the above equation, the hydraulic conductivity, $K$, is assumed to be equal in all directions (that is, isotropic) and also the same at all points in space (that is, homogenous). Further, the Dupuit assumption is applied which assumes that the flow at a point in any direction is proportional to the gradient of the water table, represented here by the variable $h$, at that location and in the given direction. The other variable, $d_{i-1/2,j}$, is the flow depth at the interface between the two cells under consideration and given by the difference between the water table and the bottom topography. Note that since the water table varies with time, $d$ is also dependent upon time. However, while we are considering the water surface $h$ at the next or unknown time level, $d$, for

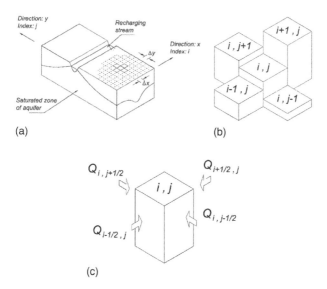

**FIGURE 5.2** (a) Seepage flow in an unconfined aquifer, approximated for solution by a two-dimensional computational grid; (b) cluster of 5 cells, used for computations; and (c) inflows to the central computational cell indexed as $i,j$ (adapted from McWhorter and Sunada, 1977).

simplicity, is considered as given or known at the current time level. That is, $d_{i,j} = h_{i,j}^t - z_{i,j}$.

The inflows from the other cells into the central cell may then be similarly written as follows:

$$Q_{i+1/2,j} = K d_{i+1/2}^t \left( h_{i+1,j} - h_{i,j} \right)^{t+\Delta t} \tag{5.15b}$$

$$Q_{i,j-1/2} = K d_{i,j-1/2}^t \left( h_{i,j-1} - h_{i,j} \right)^{t+\Delta t} \tag{5.15c}$$

$$Q_{i,j+1/2} = K d_{i,j+1/2}^t \left( h_{i,j+1} - h_{i,j} \right)^{t+\Delta t} \tag{5.15d}$$

Equations (5.15a–5.15d) together replace the first term on the right of governing Equation (5.14), which thus gives:

$$Q_{i-1/2,j} + Q_{i+1/2,j} + Q_{i,j-1/2} + Q_{i,j+1/2} = Sa^2 \left[ \frac{h_{i,j}^{t+\Delta t} - h_{i,j}^t}{\Delta t} \right] \mp R \tag{5.16}$$

The above equation may further be split into known and unknown terms as below, which helps in converting the mathematical relation easily into an equivalent computer statement.

$$
kd^t_{i-\frac{1}{2},j} h^{t+\Delta t}_{i-1,j} + kd^t_{i+\frac{1}{2},j} h^{t+\Delta t}_{i+1,j} + kd^t_{i,j-\frac{1}{2}} h^{t+\Delta t}_{i,j-1} + kd^t_{i,j+\frac{1}{2}} h^{t+\Delta t}_{i,j+1}
$$

$$
- \left[ K \left( d^t_{i-\frac{1}{2},j} + d^t_{i+\frac{1}{2},j} + d^t_{i,j-\frac{1}{2}} + d^t_{i,j+\frac{1}{2}} \right) + \frac{Sa^2}{\Delta t} \right] h^{t+\Delta t}_{i,j} = -\frac{Sa^2}{\Delta t} h^t_{i,j} \mp R \qquad (5.17)
$$

### 5.2.3 STEADY-STATE SEEPAGE BELOW FLOORS AND PILES

The governing equation for steady-state seepage flow in the vertical plane taking place below sheet piles with and without solid floors, is given by the two-dimensional continuity equation, Equation (5.8). In this section, a simplified geometry of an impervious weir floor and a pile system is considered as shown in Figure 5.3(a).

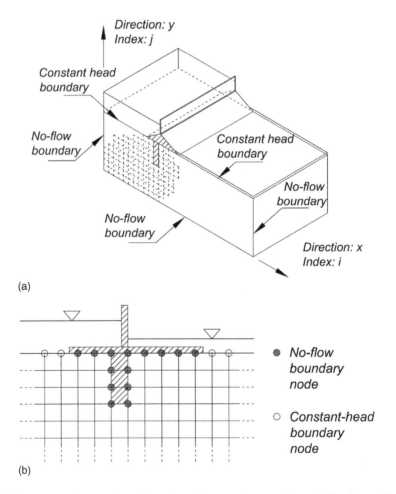

**FIGURE 5.3** (a) Computational domain and assumed rectangular grid for the problem of seepage flow below a floor and a sheet pile and (b) close-up view near the floor of the weir and the sheet pile to show the differences in the types of boundary nodes for the problem.

In geotechnical engineering, too, when a pile is driven into the ground to hold back a mass of soil behind it, the seepage flow taking place below the pile may very closely be approximated by the two-dimensional flow equation. The seepage flow below the weir floor and the sheet pile underneath is thus solved using a two-dimensional computational grid (Figure 5.3a). The grid spacing in the $x$- and $y$-directions are considered as $\Delta x$ and $\Delta y$. However, for simplicity, we consider the grid cell size in each direction to be the same (that is, $\Delta x = \Delta y$) and equal to $a$. The appropriate boundary conditions need to be specified in the computational algorithm in order to correctly reproduce the pressure head distribution within the saturated soil mass. The boundary conditions in the complete domain are indicated in Figure 5.3(a), while those close to the sheet pile and the floor of the weir are depicted in Figure 5.3(b).

The governing equation for steady-state seepage flow (Equation 5.8) may now be written in a discretized form as

$$h_{i+1,j} + h_{i-1,j} + h_{i,j+1} + h_{i,j-1} - 4h_{i,j} = 0 \tag{5.18}$$

For incorporating the constant-head boundary condition in the model, direct substitution of the values of the head would be required at the appropriate nodes. For specifying the no-flow boundary condition, the gradient of the head is specified to be zero. Accordingly, the following equation is written approximating the condition of zero-flow, where the second index $j$ and $j-1$ in the two terms correspond to the node at the boundary edge and that just adjacent to it inside the soil. The index $i$ corresponds to the node numbering when counting along the boundary edge.

$$h_{i,j-1} - h_{i,j} = 0 \tag{5.19}$$

## 5.3 PYTHON PROGRAMS

The algorithms discussed on groundwater and seepage flows in the sections above are demonstrated here through programs written in Python. The first two codes simulate the unsteady-state phenomenon in unconfined groundwater flow, while the third is a steady-state problem on seepage flow occurring below a weir and sheet pile combination due to a difference of water levels on either side of the weir. Confined groundwater flows, being simpler, are not considered.

### 5.3.1 UNSTEADY ONE-DIMENSIONAL GROUNDWATER FLOW IN AN UNCONFINED AQUIFER

The program for computing the unsteady flow in a one-dimensional unconfined aquifer is presented below. The problem assumes a horizontal initial water table corresponding to equal water levels in the two ditches on either side of the aquifer (Figure 5.4a).

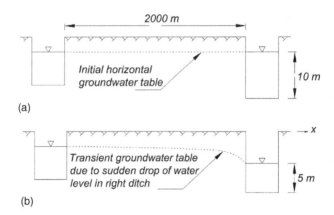

(a)

(b)

**FIGURE 5.4** One-dimensional groundwater flow between two ditches. (a) Initial condition: water level equal between two ditches and a horizontal water table profile; (b) sudden drawdown of the water level of one of the ditches, resulting in a transient water table profile.

```
# 1D transient seepage flow
import numpy as np
import matplotlib.pylab as plt

length = 2000 # Computational length
npoints = 101
Hconductivity = 10 # m per day
S = 0.1  # dimensionless
nt = 100
dt = .1 # in day
errorallow = 0.001
maxiter = 100

x = np.linspace(0, length, npoints)
dx = x[1]-x[0]

hold = np.ones(npoints)*10
hnew = np.ones(npoints)*10
c = (dx*dx*S)/(Hconductivity*dt)

A = np.zeros((npoints,npoints))
rhs = np.zeros(npoints)
time = np.zeros(nt)

# Commence time iteration over all grid points
for n in range(0,nt):
    iter = 0
    max_error = 1.0
    time[n] = time[n-1] + dt
    print("n = ",n,"time=",time[n])
```

```
while (max_error > errorallow):
    iter = iter+1
    if(iter > maxiter): break

    for i in range(1,npoints-1):
        A[i,i-1] = 2*hnew[i-1]
        A[i,i] = -4*hnew[i]-c
        A[i,i+1] = 2*hnew[i+1]
        rhs[i] = hnew[i-1]**2-2*hnew[i]**2+hnew[i+1]**2-
c*(hnew[i]-hold[i])
        A[0,0] = 1.0
        rhs[0] = hnew[0]-10
        A[npoints-1,npoints-1] = 1.0

        rhs[npoints-1] = hnew[npoints-1]-5

        delh = np.linalg.solve(A,rhs)
        max_error = np.max(abs(delh))
        hnew -= delh

    hold = hnew

    fig = plt.figure()
    ax = fig.add_subplot(1, 1, 1)
    ax.plot(x/1000,hnew,'o-')

ax.set_xlabel('Distance (km)')
ax.set_ylabel('Head (m)')
plt.show()
```

The variables used in this program are described below:

| Variable | Description | Variable | Description |
|---|---|---|---|
| length | Length of the flow domain (m) | S | Storage coefficient of the aquifer (-) |
| Hconductivity | Hydraulic conductivity (m day$^{-1}$) | nt | Number of computational time steps |
| npoints | Number of computational nodes that divides the domain (-) | dt | Time interval for computation progression (days) |

The program commences by assuming an initial condition of groundwater profile elevation at 10 m above datum. It is also assumed that immediately after the start of the simulation, the water level on the right boundary drops by 5 m. On running the program, the profile is seen to modify with time, as shown in Figures 5.5a and 5.5b, for the times 2 and 20 days after the drop of the right-boundary water level, respectively.

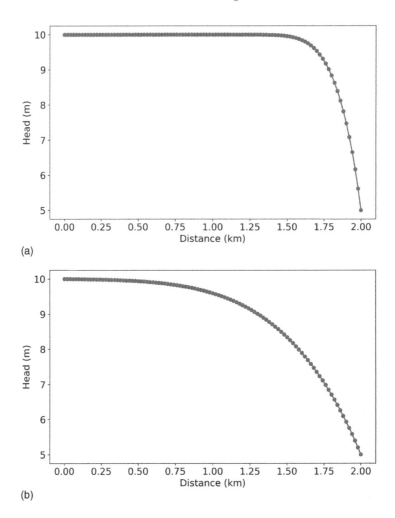

**FIGURE 5.5** Groundwater surface profile generated by a sudden drop of the right ditch water level at times: (a) 2 days from beginning and (b) 20 days from beginning. (Color image available in eBook).

### 5.3.2 UNSTEADY TWO-DIMENSIONAL GROUNDWATER FLOW IN AN UNCONFINED AQUIFER

The Python program for simulating the two-dimensional unsteady flow in an unconfined aquifer due to pumping (Figure 5.6a) is presented below. The location of the pumping wells are specified in terms of the coordinate indices $i$ and $j$ as shown in Figure 5.6b, along with the spatial dimensions of the computational domain. The number of grid nodes are also specified, with the grid size chosen as 50 m. The bottom of the aquifer may be modelled as varying in space by suitably changing the bottom-boundary elevation matrix, $z$, in the code. Presently, these data are specified as zero, meaning a horizontal impervious boundary.

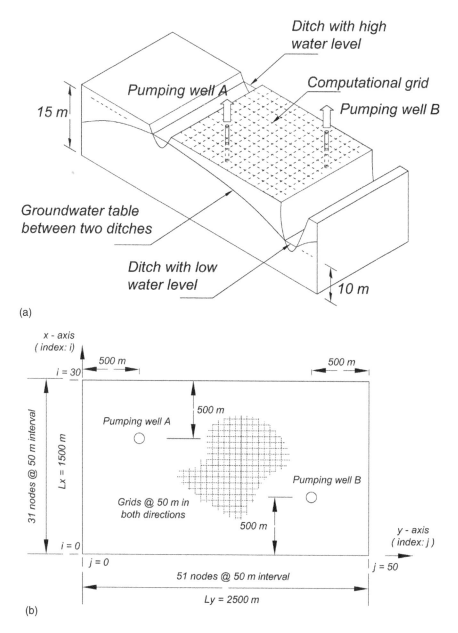

**FIGURE 5.6** Geometry and pumping details of the groundwater flow problem. (a) Perspective view of a soil block showing two ditches filled with water at unequal levels, the initial groundwater profile, an indicative computational grid, and two pumping wells, and (b) dimension of the computation domain, grid details, and location of pumping wells for the groundwater flow problem demonstrated in the Python code.

```
# Two-dmensional transient groundwater flow
import numpy as np
import matplotlib.pyplot as plt
from mpl_toolkits.mplot3d import axes3d

nx = 31
ny = 51
n = nx * ny
nt = 5000
dt = 1
a = 50.0
K = 1.0
S = 0.1
p = (S * a*a)/dt

A = np.zeros((n,n))
b = np.zeros(n)
h = np.ones((nx,ny))*15.0
z = np.zeros((nx,ny))
for i in range (0,nx):
d = h - z

# Define withdrawal rates Q as zero for all nodes except at
# nodes 12,15 and 3,4 at rates of 500 and 200 m^3/day
Q = np.zeros((nx,ny))
Q[10,40] = 10.0
Q[20,10] = 50.0

time = 0.0
for n in range(0,nt):
    time = time + dt
    print("time=",time)
    for i in range (1,nx-1):        # Speciying coefficients of
        for j in range (1,ny-1):  # matrix [A] and vector {b}
            qe = K*(d[i,j]+d[i,j+1])/2.0
            qw = K*(d[i,j]+d[i,j-1])/2.0
            qn = K*(d[i,j]+d[i+1,j])/2.0
            qs = K*(d[i,j]+d[i-1,j])/2.0
            ndiag = j + (i*ny)
            ne = ndiag + ny
            nw = ndiag - ny
            nn = ndiag + 1
            ns = ndiag - 1
            A[ndiag,ndiag] = -(qe + qw + qn + qs + p)
            A[ndiag,ne] = qe
            A[ndiag,nw] = qw
            A[ndiag,nn] = qn
            A[ndiag,ns] = qs
            b[ndiag] = Q[i,j] - p * h[i,j]
    for i in range (0,nx):  # Specifying the constant head
```

```
            for j in range (0,ny):     # Boundary Conditions
                ndiag = j + (i*ny)
                if (i == 0):
                    A[ndiag,ndiag] = 1.0 # For the left boundary
                    b[ndiag] = 10.0       # Specified head = 10m
                if (i == nx-1):
                    A[ndiag,ndiag] = 1.0 # For the Right boundary
                    b[ndiag] = 15.0       # Specified head = 15m
        for i in range(1,nx-1):          # Specifying the no-flow
            for j in range (0,ny):      # Boundary Conditions
                ndiag = j + (i*ny)
                if (j == 0):                 # For the Bottom boundary
                    A[ndiag,ndiag] = 1.0
                    A[ndiag,ndiag+2] = -1.0
                    b[ndiag] = 0.0
                if (j == ny-1):              # For the Top boundary
                    A[ndiag,ndiag] = 1.0
                    A[ndiag,ndiag-2] = -1.0
                    b[ndiag] = 0.0

    x = np.linalg.solve(A,b)
    for i in range (0,nx):
        for j in range (0,ny):
            h[i,j] = x[j + i*ny]
    d = h - z

xlist = np.linspace(0, (ny-1)*a, ny)
ylist = np.linspace(0, (nx-1)*a, nx)
X, Y = np.meshgrid(xlist, ylist)
fig = plt.figure(figsize=[10,6])
ax = fig.gca(projection='3d')

ax.plot_wireframe(X, Y, h)
ax.view_init(45,210)
ax.set_xlabel('X - axis')
ax.set_ylabel('Y - axis')
ax.set_zlabel('Head, h')
plt.show()
```

The variables used in this program are described below:

| Variable | Description | Variable | Description |
|---|---|---|---|
| nx | Computational division points in x-direction | a | Grid interval spacing, same in x- and y-directions (m) |
| ny | Computational division points in y-direction | K | Hydraulic conductivity (m day$^{-1}$) |
| nt | Number of time step intervals for simulations | S | Storage coefficient of the aquifer (-) |
| dt | Time interval for computation progression (days) | z | Matrix storing the elevation data of the bottom-boundary (m) |

The program for the two-dimensional domain, when run, computes the changing profile of groundwater table as the computation progresses with time. Figure 5.7(a) shows the profile generated by the program for a simulation time of 180 days for the given data of the aquifer, but without any pumping, while Figure 5.7(b) depicts the resulting profile for the same period of time, but in the presence of two pumping wells. Note the unequal drawdowns near the location of the two wells resulting from different withdrawal rates.

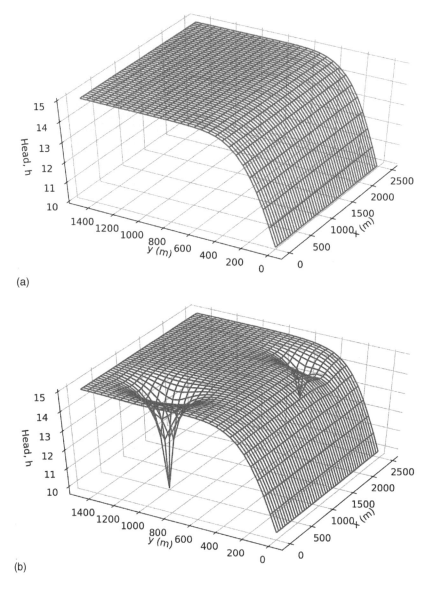

**FIGURE 5.7** Groundwater table profile between two ditches with unequal water levels plotted at the end of 180 days from start: (a) No pumping and (b) pumping at two locations. (Color image available in eBook).

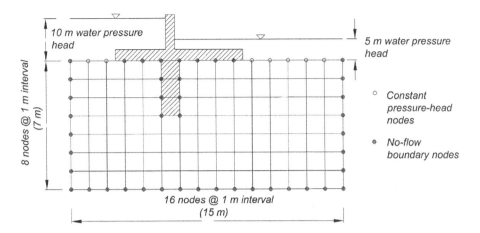

**FIGURE 5.8**   Geometry and hydraulic head data of the seepage flow occurring below a weir floor and a sheet pile with differential water levels on either side.

### 5.3.3   Steady Seepage below a Weir Floor and Sheet Pile

The Python program for simulating the two-dimensional steady flow seepage below a gated weir having sheet pile underneath is given below. The geometry of the model simulated by the program, along with the computational grid details, is shown in Figure 5.8. On either side of the floor of the weir and the gate (which extends below ground as a sheet pile), the free surface of water stands up to a depth of 10 m and 5 m, respectively, measured from a common datum.

The input data for the problem include the values of hydraulic conductivity, grid spacing, and the number of grid points in the *x* and *y*-directions, which are defined within the program. However, the specifications for the no-flow and constant-head boundaries are read into the program from an external file, `input.csv`. There are as many rows of data in this file as there are number of nodes in the discretized grid. The first few rows of the file are given below for the problem shown in Figure 5.8 and the remaining may be completed by following the sequence of the variables which, according to their order of appearance, are as follows.

```
0,0,0,-1,0,-1
0,1,0,-1,0,0
0,2,0,-1,0,0
0,3,0,-1,0,0
0,4,0,-1,0,0
0,5,0,-1,0,0
0,6,0,-1,0,0
0,7,0,-1,10,0
1,0,0,0,0,-1
1,1,0,0,0,0
1,2,0,0,0,0
1,3,0,0,0,0
1,4,0,0,0,0
1,5,0,0,0,0
```

```
1,6,0,0,0,0
1,7,0,0,10,0
...
5,0,0,0,0,-1
5,1,0,0,0,0
5,2,0,0,0,0
5,3,0,0,0,0
5,4,-1,0,0,0
5,5,-1,0,0,0
5,6,-1,0,0,0
5,7,-1,0,-1,0
...
```

The first two values in each row (separated by commas) correspond to the $i$- and $j$-grid indices of the node, followed by four neighbour or boundary codes corresponding to the nodes on the right, left, top, and bottom of the given node. The respective boundary code for a node is a negative number, if the corresponding node on that side is a no-flow boundary and a positive number if a pressure head is defined. The value of the number itself gives the value of the head in meters. If none of these conditions is met, then the neighbouring boundary code value is specified as zero. The data extracted from the data file show the values for the computational nodes on the left boundary of the grid points shown in Figure 5.8, followed by those of the next column of grid points. The sample data given at the end of the table corresponds to the computational nodes along the column of grid points that graze the left edge of the sheet pile.

Once the computations are done, and the graphs plotted on the screen, the program executes the two lines at the end that save the computed velocities in the x- and y-directions, u and v, in two separate files u_vel.csv and v_vel.csv for estimating the spread of a contaminant due to seepage in Chapter 6.

```python
# Seepage flow below floor with sheet pile - solution by the
    implicit method
import csv
import numpy as np
import matplotlib.pyplot as plt

# Input data being loaded from input.csv file to an array
    indata
indata = np.loadtxt('input.csv',delimiter=',')

k = 0.1
d = 1.0
nx = 16
ny = 8
n = nx*ny

A = np.zeros((n,n))
c = np.zeros(n)
```

```
# Finding potentials Z
for i in range (0,nx):
    for j in range (0,ny):
        ndiag = j+i*ny
        A[ndiag,ndiag] = -4.0
        r = int(indata[ndiag,2])
        l = int(indata[ndiag,3])
        t = int(indata[ndiag,4])
        b = int(indata[ndiag,5])
        nr = ndiag+ny
        nl = ndiag-ny
        nt = ndiag+1
        nb = ndiag-1

        if r < 0: A[ndiag,nl] += 1.0
        elif r == 0: A[ndiag,nr] += 1.0
        if l < 0: A[ndiag,nr] += 1.0
        elif l == 0: A[ndiag,nl] += 1.0
        if t < 0: A[ndiag,nb] += 1.0
        elif t == 0: A[ndiag,nt] += 1.0
        if b < 0: A[ndiag,nt] += 1.0
        elif b == 0: A[ndiag,nb] += 1.0
        if r > 0:
            A[ndiag,:] = 0.0
            A[ndiag,ndiag] = 1.0
            c[ndiag] = r
        if l > 0:
            A[ndiag,:] = 0.0
            A[ndiag,ndiag] = 1.0
            c[ndiag] = l
        if t > 0:
            A[ndiag,:] = 0.0
            A[ndiag,ndiag] = 1.0
            c[ndiag] = t
        if b > 0:
            A[ndiag,:] = 0.0
            A[ndiag,ndiag] = 1.0
            c[ndiag] = b

x = np.linalg.solve(A,c)

xlist = np.linspace(0, nx-1, nx)
ylist = np.linspace(0, ny-1, ny)
X, Y = np.meshgrid(xlist, ylist)

Z = np.zeros((ny,nx))
for i in range (0,nx):
    for j in range (0,ny):
        Z[j,i] = x[j+i*ny]

fig, ax = plt.subplots(figsize=(8,4))
```

```
cp = ax.contourf(X, Y, Z,
levels=[5,5.5,6,6.5,7,7.5,8,8.5,9,9.5,10])
fig.colorbar(cp)
ax.set_title('Equipotential Contours')
ax.set_xlabel('x (m)')
ax.set_ylabel('y (m)')

# Finding u and v
u = np.zeros((ny,nx))
v = np.zeros((ny,nx))
for i in range (0,nx):
    for j in range (0,ny):
        ndiag = j+i*ny
        r = indata[ndiag,2]
        l = indata[ndiag,3]
        t = indata[ndiag,4]
        b = indata[ndiag,5]
        if(r<0 or l<0):
            u[j,i] = 0
        else:
            u[j,i] = (-0.5/d)*k*(Z[j,i+1]-Z[j,i-1])
        if(t<0 or b<0):
            v[j,i] = 0
        elif(t>0):
            v[j,i] = (-1.0/d)*k*(Z[j,i]-Z[j-1,i])
        else:
            v[j,i] = (-0.5/d)*k*(Z[j+1,i]-Z[j-1,i])
fig, ax = plt.subplots(figsize=(8,4))
ax.quiver(X,Y,u,v)
ax.set_title('Velocity Vector Plot')
ax.set_xlabel('x (m)')
ax.set_ylabel('y (m)')
plt.show()

fig, ax = plt.subplots(figsize=(8,4))
ax.streamplot(X,Y,u,v)
ax.set_title('Streamline Plot')
ax.set_xlabel('x (m)')
ax.set_ylabel('y (m)')
plt.show()

np.savetxt('u_vel.csv', u, delimiter=',')
np.savetxt('v_vel.csv', v, delimiter=',')
```

The variables used in this program are described below:

| Variable | Description | Variable | Description |
|---|---|---|---|
| k | Hydraulic conductivity (m day$^{-1}$) | d | Grid interval spacing, same in x- and y-directions (m) |
| nx | Computational division points in x-direction | ny | Computational division points in y-direction |

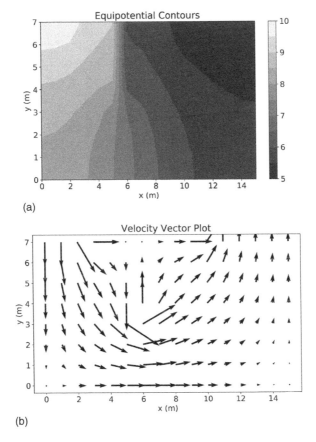

**FIGURE 5.9**  Results of the computations for seepage flow below weir floor and sheet pile: (a) equal-head or equipotential filled-contours and (b) flow vectors. (Color image available in eBook).

The program computes the pressure head at the grid points, which are plotted in Figure 5.9(a). The flow vectors at the different computational grid points are shown in Figure 5.9(b).

## REFERENCES

McWhorter, D. B. and Sunada, O. K., (1977). *Ground-water Hydrology and Hydraulics*. Water Resources Publications, Fort Collins, Colo., USA.

Wang, H. F. and Anderson, M. P. (1982). *Introduction to Groundwater Modeling: Finite Difference and Finite Element Methods*. Academic Press, London, UK.

# 6 Partial Differential Equations in Contaminant Transport

Contaminant or pollutant transport in the air, water, and soil environment is a major cause of concern to the world today. Water is one of the most potent vehicles for solute transport, which may trace its way within the soil or over the earth's surface. In addition to being advected, that is moved along with the flow of water, the contaminants may also spread laterally within the free flowing water of a river, stream, or any other water body. This is brought about by the action of diffusion due to the concentration gradient of the solute and molecular and turbulent dispersion. Different physical situations may exhibit the transport of contaminants by surface water, two of which that are commonly encountered are shown in Figure 6.1. The first figure (Figure 6.1a) depicts a flowing stream that is receiving a continuous source of pollutant at a location, which is being carried down along with the flow and also spreading across the channel's width. After some distance, the concentration of the pollutant may be assumed to have become uniformly mixed across the entire section of the stream. From thereon, the problem may be treated as a one-dimensional transport phenomenon of a completely mixed contaminant. In Figure 6.1(b), a shallow water basin is shown being subjected to an inflow containing a contaminant as a solute, which gradually spreads while progressing along with the flow in the basin. Being shallow, the concentration of the solute is assumed to be nearly uniform across the depth of water due to vertical mixing and thus the problem may be considered as occurring in the two-dimensional horizontal space within the water body.

Similarly, the water infiltrating into soils carry with them solutes from sources of pollution above ground or from sources within the ground itself. The contaminant disperses, in general in all the three directions, as it is carried along with the flow of groundwater through the voids of the soil particles. Figure 6.2(a) shows a possible case of leaching of contaminants from a source of waste dump by water which first infiltrates through the unsaturated zone of the soil and then, reaching the saturated groundwater flow below, gets conveyed along with the flow and dispersing at the same time within the soil particles. Figure 6.2(b) shows another case of a possible groundwater contamination, but for a relatively shallow aquifer flow taking place due to the head difference of water tables. The contaminant here is seen to spread mostly in the horizontal plane while being advected by the general groundwater flow. Because of the small thickness of the saturated aquifer strata, the solute is assumed to have achieved vertical uniformity in concentration of the contaminant.

DOI: 10.1201/9780429288579-6

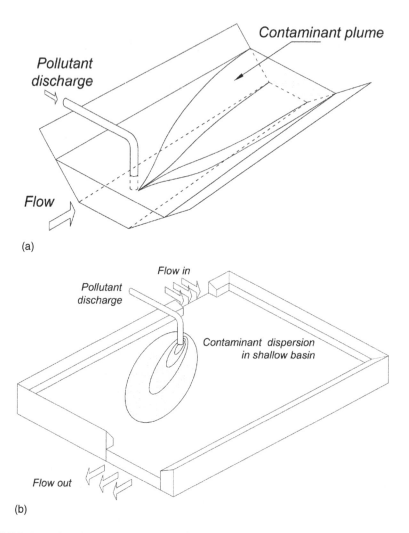

**FIGURE 6.1**   Contaminant transport in (a) one-dimensional river and (b) two-dimensional shallow water basin.

   In this chapter, we discuss the phenomena of solute transport in simplified cases of the more general situations described above occurring in surface water bodies and groundwater aquifers since the governing equations are similar – both being second-order partial differential equations. Not all solutes, however, are non-reactive and hence some constituents of the pollutants may change in concentration by getting transformed into other products. This has been seen to be more vigorous in the "hyporheic zone", which is the patch of soil at the interface of surface and groundwater domains. Hence, apart from the more apparent processes of advection and diffusion or dispersion of a solute, its reactive phase may also be taken into

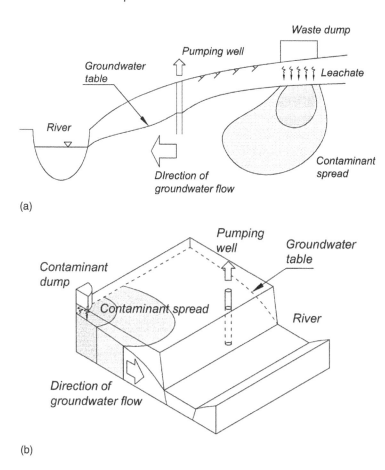

**FIGURE 6.2** Contaminant transport in (a) seepage flow in the vertical plane and (b) shallow groundwater flow in the horizontal plane.

account according to the problem being considered. In the present treatise, however, only non-reactive constituents for groundwater flows have been assumed.

## 6.1 GOVERNING EQUATIONS

The governing equations for the commonly encountered contaminant transport problems are presented in this section. We assume examples of flows from both above surface and within the ground.

### 6.1.1 GOVERNING EQUATIONS FOR REACTION-DIFFUSION, WITHOUT ADVECTION

Under a situation where there is no dominant advection of water acting as the solvent, a solute may predominantly undergo diffusion. However, if the substance is reactive and characterized by an irreversible first-order reaction so that the rate of removal of the diffusing substance is $kC$, where $k$ is a constant, the governing equation for the fate of the solute may be written as (Fetter, 1993; Hemond and Fechner-Levy, 2000):

$$\frac{\partial C}{\partial t} = D \frac{\partial^2 C}{\partial x^2} - kC \tag{6.1}$$

In Equation (6.1), $C$ is the concentration of the solute, $D$ is the diffusion coefficient, $t$ is the time variable, and $x$ the one-dimensional space variable. Thus, the concentration $C$ is seen to vary both in time and space, which is considered only in one direction in this case. The above form of equation also appears for representing the conduction of heat along a metal rod that is also losing heat from its surface at a rate proportional to its temperature. While the heat passes through the rod, it is assumed that the temperature of the rod is uniform across a section of the rod. Quite similarly, when Equation (6.1) is applied to a contaminant, it is assumed that the concentration of the solute is uniform across the cross section of the one-dimensional medium through which it is diffusing.

If the other perpendicular direction (the $y$-axis) is considered as well for diffusion in a two-dimensional plane, Equation (6.1) may be expanded as:

$$\frac{\partial C}{\partial t} = D \left( \frac{\partial^2 C}{\partial x^2} + \frac{\partial^2 C}{\partial y^2} \right) - kC \tag{6.2}$$

### 6.1.2 Governing Equations for Advection and Diffusion

The next case commonly encountered is that of the transport of a solute by the flow of a stream or by groundwater while at the same time undergoing diffusion. The governing equation for advection in one-dimension may be written as:

$$\frac{\partial C}{\partial t} + u \frac{\partial C}{\partial x} = D \frac{\partial^2 C}{\partial x^2} \tag{6.3}$$

In Equation (6.3), the terms carry their usual meanings, as defined before but with the additional term $u$, which is the flow velocity of the stream in the $x$-direction. Although the flow velocity is one-dimensional, dispersion occurs perpendicular to the flow direction as well. However, it is assumed here that the solute is completely mixed in the direction perpendicular to the flow.

When expanded to the two-dimensional case, where the flow velocity has components predominantly over a two-dimensional plane with velocities $u$ and $v$ assumed in the $x$- direction and $y$-direction, respectively, Equation (6.3) is expressed as:

$$\frac{\partial C}{\partial t} + u \frac{\partial C}{\partial x} + v \frac{\partial C}{\partial y} = D \left( \frac{\partial^2 C}{\partial x^2} + \frac{\partial^2 C}{\partial y^2} \right) \tag{6.4}$$

In Equation (6.4), it is assumed that in the direction perpendicular to the $x$-$y$ plane, the contaminant is completely dispersed, which is possible if the thickness of the medium in that direction is comparatively small.

### 6.1.3  Governing Equations for Advection, Diffusion, and Reaction

The final case that may be encountered in problems of determining the fate of a reactive solute involves all the three agents of advection by flow, diffusion, and reaction. The one-dimensional governing equation for this case is given as:

$$\frac{\partial C}{\partial t} + u\frac{\partial C}{\partial x} = D\frac{\partial^2 C}{\partial x^2} - kC \tag{6.5}$$

An equivalent two-dimensional form may be derived by expanding Equation (6.5) appropriately.

## 6.2  NUMERICAL METHODS FOR FINDING THE FATE OF A CONTAMINANT

Assessment of the fate of a contaminant influenced by dispersion and advection discussed above, the one-dimensional steady-state version, in the form of an ordinary partial differential equation, has been presented in Chapter 3. Also, the equation was solved as an initial value problem as only one boundary condition was required to be defined at any one end of the computational domain. In this chapter, the equation for the one-dimensional problem is discussed, but for the unsteady state, which now takes the form of a partial differential equation. The two-dimensional form of the unsteady equation is also discussed and applied with some applications. Further, since each of the equation contains the diffusion term defined with the double derivative of the concentration variable, they are classified as second-order differential type of equations. For a complete solution, the conditions at the boundary points of the computational domain are required to be known. In the one-dimensional case, these would be specified as either the value of the concentration of the solute (Dirichlet condition) or as some form of the gradient of the concentration (Neumann condition). In the two-dimensional case, the quantities need to be specified all along the periphery of the computational domain.

### 6.2.1  Solving the One-Dimensional Unsteady Reaction-Diffusion Problem

This is the case for diffusion of a contaminant that is also undergoing reaction, that is, by interaction with the atmosphere it is being transformed into another material at a certain rate, causing its original amount to reduce with time. Among the several choices available to solve the governing differential equation in such a case (Equation 6.1), the backward-in time central-in-space scheme of finite differences is adopted here, along with the grid shown in Figure 6.3.

The equivalent finite difference form of the governing differential equation (Equation 6.1) is presented below. In order to derive an explicit formulation, the forward finite-difference scheme is used for the time derivative while the second derivative in space, associated with the diffusion term, is approximated by a central-difference scheme.

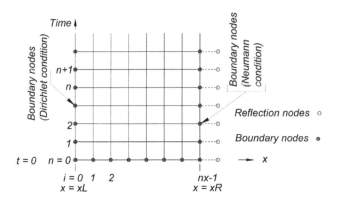

**FIGURE 6.3**  Finite difference grid used for solving the one-dimensional reaction-diffusion problem, showing typical boundary conditions at either end of the computational domain.

$$\frac{C_i^{n+1} - C_i^n}{\Delta t} = D\frac{C_{i-1}^n - 2C_i^n + C_{i+1}^n}{\Delta x^2} - kC_i^n \tag{6.6}$$

The subscript with the concentration variable $C$ stands for the space grid number, while the superscript shows the time level, with $n$ representing the current time and $n+1$, the next time step. The space grid index $i$ denotes the computational node under consideration, while $i-1$ and $i+1$ stand for the nodes just behind and just in front, respectively, of node $i$.

Equation (6.6) may be simplified by considering the non-dimensional form of the time and space step variables ($\Delta t$ and $\Delta x$, respectively) along with the diffusion coefficient $D$, as the Fourier number, $r = D\dfrac{\Delta t}{(\Delta x)^2}$. Also, by introducing the parameter $s = \dfrac{\Delta t}{K}$, finally the equation gets converted to:

$$C_i^{n+1} = rC_{i-1}^n + (1 - 2r + s)C_i^n + rC_{i+1}^n \tag{6.7}$$

For obtaining an implicit solution, the concentrations in the difference terms of the second derivative have to be expressed in terms of the values of the next time level, that is, $n+1$, instead of the present time level, $n$. This leads to the following form of the difference equation:

$$\frac{C_i^{n+1} - C_i^n}{\Delta t} = D\frac{C_{i-1}^{n+1} - 2C_i^{n+1} + C_{i+1}^{n+1}}{\Delta x^2} - kC_i^n \tag{6.8}$$

Now, the resulting equations may be written in terms of the coefficients $r$ and $s$ as:

$$-rC_{i-1}^{n+1} + (1 + 2r - s)C_i^{n+1} - rC_{i+1}^{n+1} = C_i^n \tag{6.9}$$

## 6.2.2 Solving the One-Dimensional Unsteady Advection-Diffusion Problem

Here, we consider the governing equation of the case in which there is no reactive term but an advective term is included instead (velocity, $u$, in Equation 6.3). As with the previous case of reaction-diffusion, the forward finite difference is used for the time derivative to obtain an explicit solution scheme for solving the equation. The central difference is used for approximating the second derivative associated with the diffusion term. The first derivative associated with the advection term is, however, discretized as in "upwind" schemes, such as $\dfrac{\partial C}{\partial x} = \dfrac{C_i^n - C_{i-1}^n}{\Delta x}$ if $u > 0$, or $\dfrac{\partial C}{\partial x} = \dfrac{C_{i+1}^n - C_i^n}{\Delta x}$ if $u < 0$. Here, the superscript n stands for the current time level and the subscript, denotes the computational node. The resulting discretized form of the governing equation may further be simplified by expressing some of the variables in the non-dimensional forms of $\rho = u\dfrac{\Delta t}{\Delta x}$, known as the Courant number, and $r = D\dfrac{\Delta t}{(\Delta x)^2}$, the Fourier number mentioned previously. The final form of the discretized equation may then be written (for $u > 0$) as:

$$C_i^{n+1} = C_i^n + r\left(C_{i-1}^n - 2C_i^n + C_{i+1}^n\right) - \rho\left(C_i^n + C_{i-1}^n\right) \qquad (6.10)$$

Or, equivalently as:

$$C_i^{n+1} = \left(r + \rho\right)C_{i-1}^n + \left(1 - 2r - \rho\right)C_i^n + rC_{i+1}^n \qquad (6.11)$$

For the case when $u < 0$, we have

$$C_i^{n+1} = rC_{i-1}^n + \left(1 - 2r + \rho\right)C_i^n + \left(r + \rho\right)C_{i+1}^n \qquad (6.12)$$

If an implicit solution is required, the approximations for the spatial derivatives have to be done with the values of the variables as in the next time level, $n+1$, while the time derivative remains as before.

## 6.2.3 Solving the One-Dimensional Combined Unsteady Advection, Diffusion, and Reaction Equation

The most general one-dimensional case for contaminant transport, represented by Equation 6.5, including all the terms of advection, diffusion, and reaction, may be obtained by combining methods demonstrated in the previous sections. Thus, assuming $u > 0$, the explicit form of the discretized equation may read as:

$$\frac{C_i^{n+1} - C_i^n}{\Delta t} + \frac{u}{\Delta x}\left(C_i^n - C_{i-1}^n\right) = \frac{D}{\Delta x^2}\left(C_{i-1}^n - 2C_i^n + C_{i+1}^n\right) - kC_i^n \qquad (6.13)$$

The above equation may be converted to an implicit scheme by replacing the concentration values in the two spatial derivative terms with those of the next time level, $n+1$.

### 6.2.4 SOLVING THE TWO-DIMENSIONAL UNSTEADY ADVECTION AND DIFFUSION EQUATION

The two-dimensional case of contaminant transport by advection and diffusion, characterized by Equation 6.4, may be discretized in a way that is similar to the one-dimensional equations. However, the grid discretization in space in this case should be in two dimensions, while the time discretization remains the quite same as shown in Figure 6.4.

For an explicit scheme of finite difference discretization, the following time and space difference formulations are used for approximating the corresponding derivatives terms:

Time derivative approximated as: $\dfrac{\partial C}{\partial t} = \dfrac{C_{i,j}^{n+1} - C_{i,j}^{n}}{\Delta t}$

The first derivatives of the advection terms, considering the velocities $u$ and $v$ in the $x$- and $y$-coordinate directions as positive, are approximated as:

$$u\frac{\partial C}{\partial x} + v\frac{\partial C}{\partial y} = u\frac{C_{i,j}^{n} - C_{i-1,j}^{n}}{\Delta x} + v\frac{C_{i,j}^{n} - C_{i,j-1}^{n}}{\Delta y}$$

And, finally, the second derivatives of the diffusion terms are expressed in difference terms as: $D\left(\dfrac{\partial^2 C}{\partial x^2} + \dfrac{\partial^2 C}{\partial y^2}\right) = D\left(\dfrac{C_{i-1,j}^{n} - 2C_{i,j}^{n} + C_{i+1,j}^{n}}{\Delta x^2} + \dfrac{C_{i,j-1}^{n} - 2C_{i,j}^{n} + C_{i,j+1}^{n}}{\Delta y^2}\right)$

Considering the grid points in the x- and y-directions to be spaced equally at a distance of $a$, that is, $\Delta x = \Delta y = a$, the complete discretized equation may be expressed as given below by substituting the above approximations into the governing equation (Equation 6.4).

$$\frac{C_{i,j}^{n+1} - C_{i,j}^{n}}{\Delta t} + u\frac{C_{i,j}^{n} - C_{i-1,j}^{n}}{\Delta x} + v\frac{C_{i,j}^{n} - C_{i,j-1}^{n}}{\Delta y} = \frac{D}{a^2}\left[\begin{array}{l}\left(C_{i-1,j}^{n} - 2C_{i-j}^{n} + C_{i+1,j}^{n}\right) + \\ \left(C_{i,j-1}^{n} - 2C_{i-j}^{n} + C_{i,j+1}^{n}\right)\end{array}\right] \quad (6.14)$$

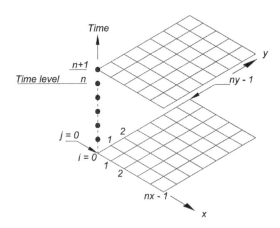

**FIGURE 6.4** Finite difference grid used for the two-dimensional advection-diffusion problem.

The above explicit formulations are seen to hold good for the following condition:

$$\Delta t \le \left[ \frac{u}{\Delta x} + \frac{v}{\Delta y} + D \left( \frac{1}{\Delta x^2} + \frac{1}{\Delta y^2} \right) \right]^{-1} \tag{6.15}$$

If one or both the velocities, $u$ and $v$, are negative, the approximations for the derivatives in the advective terms are required to be written as $u \dfrac{\partial C}{\partial x} = u \dfrac{C_{i+1,j}^n - C_{i,j}^n}{a}$, if $u < 0$, and $v \dfrac{\partial C}{\partial x} = v \dfrac{C_{i,j+1}^n - C_{i,j}^n}{a}$, if $v < 0$. This method is sometimes referred to as upwinding, that is, giving more weight to the velocities on the upstream (equivalent to the upwind direction for air flows) of the point of concern.

## 6.3   PYTHON PROGRAMS

In this section, the processes for contaminant transport in one- and two dimensions by surface and subsurface flows are demonstrated using representative programs in Python which make use of the numerical algorithms explained in the preceding section.

### 6.3.1   ONE-DIMENSIONAL UNSTEADY REACTION-DIFFUSION PROBLEM

A program in Python for the one-dimensional unsteady reaction-diffusion problem is presented below. The data specified in the code assume a 10-unit long water body, divided into segments by 20 computational grid points or nodes. The initial and final times are specified, also divided into 20 time steps. Two central nodes are specified with known concentrations to start with, which diffuse gradually with time. Two different boundary conditions are specified at the two ends of the computation domain for all times: at the left boundary, the concentration is specified as equal to zero; at the right boundary a gradient condition is specified. Assumed values of diffusion and reaction rates are input to the program as data values.

```
# Program for solving the diffusion-reaction problem in
   one-dimension
import numpy as np
from mpl_toolkits.mplot3d import Axes3D
import matplotlib.pyplot as plt
from matplotlib import cm

nx = 20 # Number of space grid points
nt = 20 # Number of tme grid points

xL =  0.0 # left boundary coordinate
xR = 10.0 # Right boundary coordinate
dx = (xR - xL)/(nx - 1) # Space discretization step size

t_start = 0.0 # Initial time
```

```
t_end = 2.0 # Final time
dt = (t_end - t_start)/(nt - 1) # Time discretization step size

D = 1.0        # Diffusion coefficient
K = -2.0 # Reaction rate
g = 0.5 # Specified gradient at left boundary

r = dt*D/dx**2
s = dt*K;

C = np.zeros((nx+1,nt)) # Intializing the U matrix with all
   zeros
C[int(nx/2),0] = 1.0 # Initial concentration at central node
C[int(nx/2)+1,0] = 1.0 # Initial concentration at central + 1
  node

A = np.zeros((nx+1,nx+1)) # Intializing coefficient matrix
b = np.zeros(nx+1)        # Intializing right hand side vector
C[0,:] = 0.0              # Specifying left boundary condition
# Gradient condition is specified at right boundary: du_dx = g

for j in range (1,nt):
    for i in range (1,nx):
        ndiag = i
        A[ndiag,ndiag] = 1.0 + 2.0*r - s
        A[ndiag,ndiag-1] = -r
        A[ndiag,ndiag+1] = -r
        b[ndiag] = C[ndiag,j-1]
    A[0,0] = 1.0
    b[0] = C[0,j]
    A[nx,nx] = 1.0
    A[nx,nx-2] = -1.0
    b[nx] = 2.0 * dx * g
    x = np.linalg.solve(A,b) # Solving the unknowns
    C[:,j]=x # Assigning the unknowns to appropriate U values

# Surface plotting of concentration in space and time

xspan = np.linspace(xL, xR, nx+1)
tspan = np.linspace(t_start, t_end, nt)
X, T = np.meshgrid(tspan, xspan)
fig = plt.figure()

ax = fig.gca(projection='3d')
surf = ax.plot_surface(X, T, C, cmap=cm.coolwarm,
antialiased=False)
ax.set_xticks([0, 0.5, 1.0, 1.5, 2.0])
ax.set_xlabel('Time')
ax.set_ylabel('Space')
ax.set_zlabel('C')
plt.tight_layout()
plt.show()
```

The variables used in this program are described below:

| Variable | Description | Variable | Description |
|---|---|---|---|
| nx | Number of space grid points | D | Diffusion coefficient $(m^2 s^{-1})$ |
| nt | Number of grid points in time | nt | Reaction rate $(s^{-1})$ |
| xL | Left boundary coordinate (m) | xR | Right boundary coordinate (m) |
| t_start | Initial or starting time (s) | t_end | Final or ending time (s) |
| grad | Specified concentration gradient at left boundary | | |

On executing the program, the variation of the concentration of the solute in the assumed one-dimensional water body is computed along its length, and through time. The changes in concentration with time and length are then graphed as a composite three-dimensional plot, as shown in Figure 6.5.

## 6.3.2 ONE-DIMENSIONAL UNSTEADY ADVECTION-DIFFUSION PROBLEM

A program in Python for computing the variation of concentration in a one-dimensional channel under the advection and diffusion processes is presented below. The channel is the same as that for finding the gradually varied flow (GVF) profile behind a dam, Section 3.3.3, having a length of 20,000 m. The velocities (in ms⁻¹) computed by the program, and stored in a .csv file by the name "GVF.v.csv" at 101 computational points, have been imported in the present code as inputs. Although the velocities do not change with time, the problem simulated by the code assumes that at the upstream end of the channel, a contaminant source (provided as a concentration of

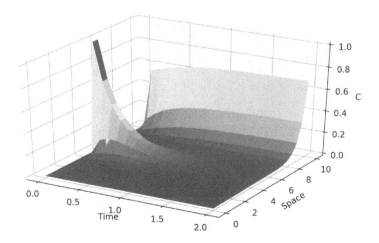

**FIGURE 6.5** A composite plot showing the variation of concentration (C) in a one-dimensional domain under reaction and diffusion processes with time and space. (Color image available in eBook).

50 units of the solute) remains active from the start up to a time of 4800 s beyond which, the source of contaminant is stopped. The code is time stepped at intervals of 60 s until the final simulation time of 28,800 s is reached. For carrying out the calculations, a diffusion coefficient of 50 units is assumed.

```python
# Pollutant transport (unsteady) in backwater GVF flow behind
# a dam
import numpy as np
import matplotlib.pylab as plt

length = 20000
npoints = 21
dt = 60
time_simulation = 28800
nt = int(time_simulation/dt)
Csource = 2.0
Ctime = 4800
D = 50

x = np.linspace(0, length, npoints)
dx = x[1]-x[0]
dtdx = dt/dx
v = np.loadtxt('GVF.v.csv',delimiter=',')
print (v)

Cold = np.zeros(npoints)
Cnew = np.zeros(npoints)
Cinlet = np.zeros(nt)
Cquarter = np.zeros(nt)
Cmidpoint = np.zeros(nt)
time = np.zeros(nt)

time[0]=0.0
for n in range (1,nt):
    time[n] = time[n-1] + dt
    if(time[n] < Ctime):
        Cold[0] = Csource
    else:
        Cold[0] = 0.0
    for i in range(1,npoints-1):
        vavg = 0.5*(v[i-1]+v[i])
        Courant = (vavg*dtdx)
        Fourier = (D*dt/(dx*dx))
        Cnew[i] = (Courant+Fourier)*Cold[i-1]+(1-2*Fourier-
Courant)*Cold[i]\
        +Fourier*Cold[i+1]
    Cnew[npoints-1] = Courant*Cold[npoints-2]+(1-
Courant)*Cold[npoints-1]
    Cold = Cnew
    Cinlet[n] = Cnew[0]
    Cquarter[n] = Cnew[int(npoints/4)]
    Cmidpoint[n] = Cnew[int(npoints/2)]

fig = plt.figure()
```

```
ax = fig.add_subplot(1, 1, 1)
ax.plot(time, Cinlet, label='Inlet')
ax.plot(time, Cquarter, label='L/4 point')
ax.plot(time, Cmidpoint, label='L/2 point')
ax.set_xlabel('Time (s)')
ax.set_ylabel('Concentration (mg/l)')
plt.legend()
plt.show()
```

The variables used in this program are described below:

| Variable | Description | Variable | Description |
|---|---|---|---|
| length | Length of the channel: same as that used in the GVF problem (m) | time_simulation | Total time of simulation (s) |
| npoints | Number of computational nodes that divides the domain (-) | nt | Number of computational time steps (m) |
| Csource | Concentration of the contaminant source (mg $l^{-1}$) | dt | Time interval for computation (s) |
| Ctime | Time up to which the contaminant is injected (s) | D | Diffusion coefficient ($m^2 s^{-1}$) |

As mentioned, the program uses the velocities at each grid points from the previously computed values of the GVF profile problem. The source of contaminant is assumed to be located at the upstream end of the channel and the pollutant is injected only up to a certain point in time, thus making the solute concentration vary with time and along the length of the channel. On running the program, the variations of the concentration are plotted at representative locations along the channel, as functions of time, as shown in Figure 6.6.

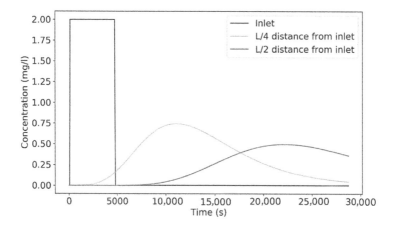

**FIGURE 6.6**  Variation in concentration with time in the GVF backwater flow behind a dam at the locations (a) upstream end, source of the injected pollutant, (b) quarter length of the channel from the upstream, and (c) mid-length of the channel. (Color image available in eBook).

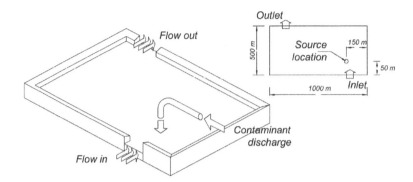

**FIGURE 6.7** Schematic of a shallow water basin with a pollutant discharge source. The basin has the same geometry as in Figure 4.10b. The image inset indicates the location of the contaminant source with respect to the basin geometry.

### 6.3.3 TWO-DIMENSIONAL UNSTEADY ADVECTION-DIFFUSION PROBLEM

A program for computing the variation of the concentration of a contaminant in a two-dimensional shallow water basin, discussed previously in Section 4.2.6, is presented in this section. The basin, assumed to be flat-bottomed and having one inlet source and one outlet or exit point, was analysed for the circulation velocities in the x- and y-directions at all computational grid points using the Python program presented in Section 4.3.6. Here, the same basin is considered for predicting the dispersion of a pollutant that is assumed to be released into it at a given location (Figure 6.7). Since the velocities at the grid points were already obtained by the above code and saved in files, these are imported and made use of here while carrying out the computations of solute transport. It may be recalled that advection of the solute occurs due to the velocity of the flowing water, while its diffusion takes place because of the concentration gradient of the solute existing between adjacent points. For the sake of simplicity, we shall assume a uniform diffusion coefficient throughout the basin.

The program in Python, solving Equation (6.4), assumes that the contaminant is conservative and thus no reaction process takes place. The code also specifies the following data: concentration of the solute at the injection point, grid indices specifying the location of the injection point, an assumed diffusion coefficient, grid spacing, computational time step, number of grid points or computational nodes in the x- and y-directions, and the total time of simulation. Since the water basin is shallow, it is assumed that the variation of concentration of the solute with depth is negligible. This implies that the concentration of the solute within the basin may be assumed to vary only with time and over the horizontal extents.

```
# Program for solving the two-dimensional contaminant
  dispersion problem
import csv
import math
import numpy as np
from mpl_toolkits.mplot3d import Axes3D
import matplotlib.pyplot as plt
```

```
Csource=10.0
ic = 85
jc = 2
D = 5
a=10
dt=1
nx=101
ny=51

timesimulation = 1200
nt = int(timesimulation/dt)
u = np.loadtxt('u.csv',delimiter=',')
v = np.loadtxt('v.csv',delimiter=',')

time=0.0
conc=np.zeros((nx,ny))

# Main program
for n in range (1,nt+1):
    time = time + dt

    for i in range(1,nx-1):
        for j in range(1,ny-1):
            if(i==ic and j==jc):
                conc[i,j]=Csource
            else:
                ucell=u[i,j]
                vcell=v[i,j]
                term1=conc[i,j]/dt
                if(ucell > 0):
                    term2=ucell/a*(conc[i,j]-conc[i-1,j])
                else:
                    term2=ucell/a*(conc[i+1,j]-conc[i,j])
                if(vcell > 0):
                    term3=vcell/a*(conc[i,j]-conc[i,j-1])
                else:
                    term3=vcell/a*(conc[i,j+1]-conc[i,j])

                term4=D/a**2*(conc[i+1,j]+conc[i-1,j]\
                        +conc[i,j+1]+conc[i,j-1]-4*conc[i,j])

                conc[i,j]=dt*(term1-term2-term3+term4)
x=np.mgrid[0:nx]*a
y=np.mgrid[0:ny]*a
[Y,X] = np.meshgrid(y,x)

fig,ax=plt.subplots(figsize=(8,4))
cp = ax.contourf(X, Y, conc)
#fig.colorbar(cp) # Add a colorbar to a plot
ax.set_title('Concentration Contour Plot')
ax.set_xlabel('x (m)')
```

```
ax.set_ylabel('y (m)')
plt.show()
```

The variables used in this program are described below:

| Variable | Description | Variable | Description |
|---|---|---|---|
| Csource | Concentration of the contaminant source (mg l⁻¹) | nx | Discretization points in the x-direction (i-index) |
| ic | i-coordinate location of the pollutant source | ny | Discretization points in the y-direction (j-index) |
| jc | j-coordinate location of the pollutant source | nt | Number of computational time steps (m) |
| D | Diffusion coefficient (m²s⁻¹) | timesimulation | Total time up to which simulation is carried up to (s) |
|  |  | dt | Time interval for computation (s) |

The Python code computes the solute concentration at all grid points and at all time steps, till the end-time for the simulations is reached. For the given contaminant source, located at grid coordinates ic = 45 and jc = 85, the variations in space of the solute concentration at times $t = 120$ s and 3600 s, are shown in Figures 6.8 (a) and (b), respectively.

### 6.3.4  CONTAMINANT DISPERSION FOR SEEPAGE BELOW SHEET PILE AND FLOOR

In this section, a Python program is presented for computing the dispersion of contaminant injected into the steady-state seepage flow occurring through the soil below an impervious floor attached to a sheet pile. This has relevance to a similar problem that was discussed in Chapter 5, in which the seepage flow velocities for the given boundary conditions were computed. For demonstration, the data of the same example (geometry specified in Figure 5.8) are used here. The seepage flow velocities in the x- and y-directions, computed for each grid point and saved in separate files by the program (Section 5.3.3), are imported by the present code. The relevant governing equation, Equation (6.4), when solved along with known seepage velocities and a specified diffusion coefficient, computes the spread of the contaminant plume within the soil. Although the form of the governing equation for contaminant transport in soils is similar to that for surface flows, the diffusion coefficient for soil contaminant transport has to be estimated from several factors, the details of which may be obtained from advanced texts, like Fetter (1993) or Pinder and Celia (2006). In the code presented in this section, a simplified equation for the transport of contaminant in soils has been used, for the sake of demonstration and application of Python programming in such cases.

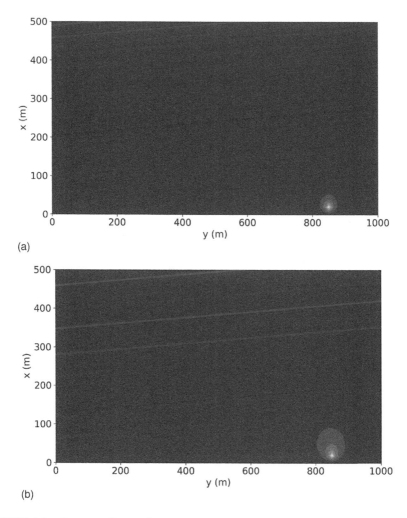

**FIGURE 6.8** Concentration profiles generated by a continuous release of pollutant source, computed for (a) 120 s and (b) 3600 s. (Color image available in eBook).

Recall that, for the computation of seepage flow velocities in soils, the presence of the solid floor and the sheet pile require some of the computational grid nodes to be treated as no-flow boundaries (Section 5.3.3). The data file containing the related information for the specification of these nodes, therefore, is also imported to the present program. The location coordinates from where the contaminant enters the soil has to be separately defined in the code.

```
# Contaminant transport in seepage
import csv
import numpy as np
import matplotlib.pyplot as plt

# Input data;
```

```
Csource=100.0
D = 50.0
ic = 4
jc = 3
a = 100
dt = 60
nx = 16
ny = 8

time_simulation = 2400
nt = int(time_simulation/dt)
print(nt)

indata = np.loadtxt('input02.csv',delimiter=',')
u = np.loadtxt('u_vel.csv',delimiter=',')
v = np.loadtxt('v_vel.csv',delimiter=',')

#u=u
# v=-v
#u=np.zeros((ny,nx))
#v=np.zeros((ny,nx))

nrows = len(u)
ncols = len(u[0])
print(nrows,ncols)

#A = np.zeros((n,n))
#c = np.zeros(n)

time = 0.0
conc = np.zeros((ny,nx))
# Main program
for n in range (1,nt+1):
    time = time + dt

    for i in range (0,nx):
        for j in range (0,ny):
            n = j+i*ny
            term_cell = (-(u[j,i]+v[j,i])/a-(4*D/a**2))
                        *conc[j,i]
            r = int(indata[n,2])
            l = int(indata[n,3])
            t = int(indata[n,4])
            b = int(indata[n,5])
            ucell=u[j,i]
            vcell=v[j,i]

            if (r!=0 or l!=0):
                term_left = 0
                term_rght = 0
            else:
                if (ucell>0):
```

```
                    term_left = (ucell/a+D/a**2)*conc[j,i-1]
                    term_rght = (D/a**2)*conc[j,i+1]
                else:
                    term_left = (D/a**2)*conc[j,i-1]
            if (t!=0 or b!=0):
                term_bot = 0
                term_top = 0
            else:
                if (vcell>0):
                    term_bot = (vcell/a+D/a**2)*conc[j-1,i]
                    term_top = (D/a**2)*conc[j+1,i]
                else:
                    term_bot = (d/a**2)*conc[j-1,i]
                    term_top = (vcell/a+D/a**2)*conc[j+1,i]
            conc[j,i] += dt*(term_cell+term_left+term_rght
                             +term_bot+term_top)
            if(i==ic and j==jc):
                conc[j,i] += Csource

xlist = np.linspace(0.0, nx-1, nx)
ylist = np.linspace(0.0, ny-1, ny)
X, Y = np.meshgrid(xlist, ylist)

fig,ax=plt.subplots(figsize=(8,4))
cp = ax.contourf(X, Y, conc)
fig.colorbar(cp) # Add a colorbar to a plot
ax.set_title('Concentration Contour Plot (SeepContaminant)')
ax.set_xlabel('x (m)')
ax.set_ylabel('y (m)')
plt.show()
```

The variables used in this program are described below:

| Variable | Description | Variable | Description |
|---|---|---|---|
| Csource | Concentration of the contaminant source (mg l$^{-1}$) | nx | Discretization points in the x-direction (i-index) |
| ic | i-coordinate location of the pollutant source | ny | Discretization points in the y-direction (j-index) |
| jc | j-coordinate location of the pollutant source | nt | Number of computational time steps (m) |
| D | Diffusion coefficient (m$^2$s$^{-1}$) | time_ simulation | Total time up to which simulation is carried up to (s) |
| | | dt | Time interval for computation (s) |

The Python code, when run, displays the variation in concentration of contaminants within the soil by diffusion and advection at the end of a specified simulation time. The program may be used to check the spread of a contaminant in soil for

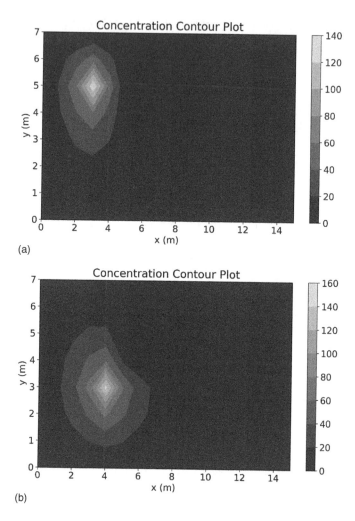

**FIGURE 6.9** Concentration profiles generated by a continuous release of pollutant source, computed for (a) $ic = 3$ and $jc = 5$ and (b) $ic = 4$ and $jc = 3$, after 2400s. (Color image available in eBook).

different source locations by varying the coordinates, `ic` and `jc`. Figures 6.9(a) and (b) show the spread of contaminants for source coordinates `ic = 3` and `jc = 5`, and `ic = 4` and `jc = 3`, respectively, at the end of a simulation period of 2400 s.

## REFERENCES

Fetter, C. W. (1993).*Contaminant Hydrogeology*. Macmillan Publishing Company, New York, USA.

Hemond, H. F. and Fechner-Levy, E. J.. (2000) *Chemical Fate and Transport in the Environment*. Academic Press, 2nd edition, San Diego, USA.

Pinder, G. F. and Celia, M. A. (2006). *Subsurface Hydrology*. John Wiley & Sons, New Jersey, USA.

# 7 Simple Data-Based Models

In the previous chapters, mathematical expressions representing different physical processes of hydrological sciences and water resources engineering were described and techniques to solve them under specified constraints and conditions were explained through computer programs written in Python. Some of these equations were algebraic functions expressed in terms of one or more variables, while others were expressed in the form of ordinary or partial differential equations (ODEs and PDEs). In this chapter, however, we shall explore some mathematical and statistical techniques that may be useful when a set of observed data is the only available information for a given process and some meaningful set of inferences is required to be extracted from it. However, for the physical processes connected to the data sets, the governing mathematical equations may not be readily available or probably too complicated to solve. We, therefore, take help of alternate data-based techniques, as discussed in this chapter, for obtaining the desired inferences out of the data.

It is possible that the observed data, which are dependent variables, may be known at discrete points of time. Here, time is the independent variable that is counted from a definite instant. On the other hand, the data values could also be measured at different locations, with reference to an origin in space, but all at the same time or over a short interval of time. For these data, the space coordinates – either in one- or two dimensions are the independent variables. Processes that are more complex may be recorded as data values at different locations or space points, and also at different time levels. These are the more general spatio-temporal data, dependent upon both space and time. However, we shall not consider this general case but limit ourselves to solely spatial or solely temporal cases which may be handled with relatively basic or elementary data-analysing techniques. These include the data values obtained for atmospheric, hydrologic, and hydraulic parameters measured along one linear direction at several spatial coordinates at a given instant of time or at one spatial location measured at different instants of time. The examples chosen for demonstration in this chapter are commonly encountered in hydrometeorological projects for undergraduate and graduate studies. The methods of analysis discussed here are also indicative of the general approaches in data analysis. For advanced analysis techniques, the reader is referred to publications such as Martinson (2018).

Some instances of data that are dependent upon time include the following: (i) water level of a groundwater observation well measured once a week or every day; (ii) discharge of a river at a gauging station measured each day; (iii) temperature observed at a meteorological station every hour; (iv) water levels of a tidal river recorded at a station every quarter of an hour; or (v) infiltration rates in a field experiment observed at intervals of 5 or 10 minutes. Similarly, observed values corresponding to a spatial dimension include examples like (i) salinity measured at different depths in a lake at a given location; (ii) soil moisture measured at different depths of soil in a field; or (iii) temperature and pressure of the air measured at different elevations above the earth surface by a meteorological balloon released from an observation station at a certain time of the day. This chapter presents sample data similar to those discussed above and examines the mathematical methods suitable for their analyses. Finally, sample programs in Python are presented demonstrating the application of the techniques.

## 7.1  ENVIRONMENTAL DATA AND MOTIVATION FOR DATA ANALYSIS

Although many different types of environmental data are measured for scientific and engineering analysis, the following two types are discussed herein, as mentioned above: (a) data values that are recorded with time, and (b) those that are recorded at locations along a direction in space, measured from a reference point. In addition, an example is demonstrated that requires values extracted from a map. Such maps could be, for example, a topographic survey map showing equal elevation contours, or a rainfall intensity map plotting iso-pluvial or equal rainfall-intensity contours.

### 7.1.1  TIME-SERIES DATA: VARIATIONS IN TIME

Certain types of data in water resources engineering and environmental hydrology and hydraulics are available as values recorded over time. There is usually a fixed interval of time between two observations. The set of data forms what is known as a "time-series". A few examples of these are discussed here in the context of scientific observations in the above fields.

#### 7.1.1.1  Hourly Record of Temperature and Humidity

The hydro-meteorological variables of temperature and relative humidity recorded at a weather station at every hour each day are typical examples of time-series data. When the two variables are plotted as discrete data points against the specific times at which they were recorded, the resulting graphical plot looks similar to that shown in Figure 7.1.

There are 24 data points for each day and thus for the 7 days, there are a total of 168 points for each of the two parameters in Figure 7.1. The data represented graphically help us to understand the fluctuations of the temperature and humidity throughout the day and over the week. It may be of interest to know, for example, the value of these variables in between two respective data points, and a mathematical representation of the data values would be the best option for answering these questions precisely. The graphs of Figure 7.1 also indicate, albeit intuitively, that there appears to be an inverse relation between temperature and humidity since within each day as the temperature rises, the humidity reduces and vice versa. However, here too, a

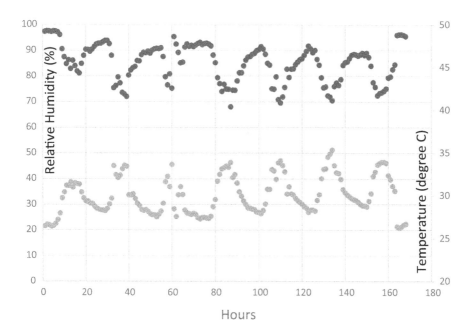

**FIGURE 7.1**   Time plot of temperature (lower graph), and relative humidity (upper graph), recorded at the Alipore Meteorological station in Kolkata, India, between July 14 and 20, 2020. (Color image available in eBook).

mathematical relation is required for expressing this negative correlation between the variables in quantitative terms. In Section 7.2, the mathematical techniques by which some of the above questions may be answered are discussed. Some other observations from the graphs are also possible which, though important, are not discussed here. For example, the average temperature each day clearly seems to increase over the week while the average daily humidity appears to fall over the same period and a quantified analysis may be possible with techniques on trend analysis. . Again, a cyclic trend in the temperature and humidity data variables is visually discernible, varying with the time of the day. This trend may be analysed quantitatively for its periodicity by other important techniques. However, these and other advanced methods on data analysis are not discussed in this book and the interested reader may refer to publications such as Martinson (2018).

### 7.1.1.2   Record of Daily River Stage and Discharge

At a river gauging station, the measurement of discharge by field methods is generally carried out only once each day. This is because the use of a current meter for velocity measurement and a depth sounder or an Acoustic Doppler Current Profiler, if available, for measuring the discharge is rather time consuming, especially when the river is wide and rough. The average river stage or its surface water level measured above a datum during the time it takes to conduct the measurements is also observed simultaneously. Thus, a time-series record is generated that stores the average daily discharge and average water level at a river gauging station, for an entire year or any other duration of time. A typical data set of stage and discharge, one pair for a day, is plotted in Figure 7.2 for a typical river gauging station of the United States Geological Survey (USGS).

For assessing the maximum discharge that passes through a river gauging station in a day and especially to know the peak discharge over the total duration, say, of a flood event, the above data do not give a ready answer. This is because both the recorded discharge and water level correspond to a specific time each day and the peak discharge may not have been captured during the daily observations. A solution to this problem could be the measuring of the flow values at closer intervals of time, such as at every quarter of an hour, or at each full hour every day. However, it is not possible to practically achieve this because of the reasons mentioned earlier. Nevertheless, measuring the level of water at shorter intervals of time is relatively easy since this can be done with the help of a gauging staff or an ultrasonic level recorder and is much less complicated and time consuming than measuring the discharge. An estimate for the discharge may then be obtained from the measured value of a given water stage with the help of a correlation formula connecting the two variables. It may be observed from Figure 7.2 that the water level and discharge at the measuring point rise and fall almost simultaneously, indicating that a positive correlation does exist between the two variables. Figure 7.2 contains 61 pairs of data

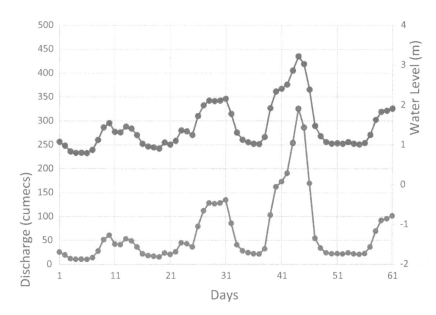

**FIGURE 7.2**  Daily discharges (lower graph) and water levels (upper graph) recorded for the Appomattox River at Matoaca, VA, USA, from October 5, to December 4, 2020. Sourced from the website https://waterdata.usgs.gov/nwis/rt. (Color image available in eBook).

points, with one pair of data pertaining to a day each and when these are plotted as in Figure 7.3, a clear pattern relating the two variables is seen to emerge.

Figure 7.4 presents a typical time plot of river stage, read by the USGS stream gauge sensors recording at 15-minute interval, for the same station, the daily data of which are plotted in Figure 7.2. These short-interval stage or water level values of Figure 7.4 may be converted to a corresponding discharge graph, also at the same short interval of time, if the correlation equation relating the water stage and discharge variables from the data plot of Figure 7.3 is known. In this chapter, we shall discuss the mathematical tools that are used in deriving a relation between the two variables forming the data set, which will help in converting the water stage values of Figure 7.4 into corresponding values of discharges. Needless to mention, this would in turn help in assessing the peak flood discharge which might have been missed during the daily exercise of river discharge measurement.

### 7.1.1.3  Variation of the Rate of Infiltration in Soil with Time

Infiltration of water into soil is usually measured with a double-ring infiltrometer. The rate of infiltration, when plotted as a function of time for a typical example, appears as shown in Figure 7.5. The infiltration rate (denoted by the variable $i$) is initially high when the soil is relatively dry, but decreases rapidly with time as the soil gets saturated. The measurements are thus taken at closer intervals of time at the

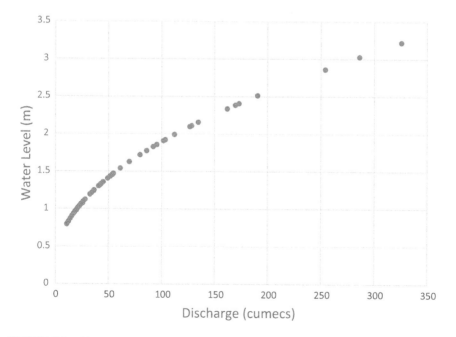

**FIGURE 7.3**  River stage versus discharge plotted for the data of Figure 7.2. The best-fitting curve through the data points, called a stage-discharge curve, helps in converting known water levels to discharge values. (Color image available in eBook).

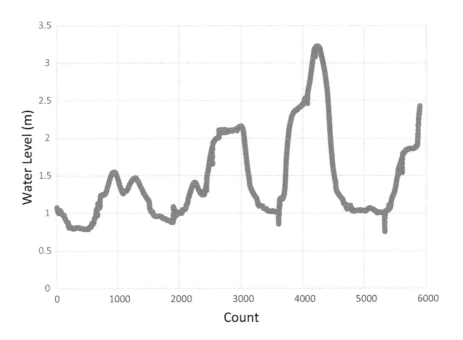

**FIGURE 7.4** Stage (water level) observed for the Appomattox River at Matoaca, VA, USA, at 15-minute interval from October 5 to December 4, 2020. The number of data points is nearly 5900; a few data points were unrecorded or misread due temporary machine malfunctioning. (Color image available in eBook).

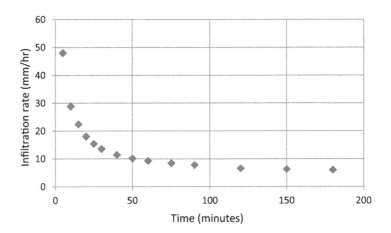

**FIGURE 7.5** Rate of infiltration, plotted against time for a typical example (sourced from Adindu et al., 2013). (Color image available in eBook).

beginning of the experiment. As time progresses, "$i$" approaches a constant value and then the time interval between the observations is increased. This example shows that data may be recorded at irregular intervals of time as well. However, the purpose of these data is not to analyse a time-series but rather to find the equilibrium rate of infiltration. Hence, although being measured as a function of time, the purpose of the data is not the same as those for the examples on time-series given above. More details on measuring infiltration in soils may be obtained from the Food and Agricultural Association (FAO) web document (http://www.fao.org/3/s8684e/s8684e0a.htm).

## 7.1.2 Data Recorded in One-Dimensional Space

Some data in environmental and hydraulic sciences, and water resources engineering are observed at fixed distances measured from an origin, but all at a specific time. Some of such examples are given below.

### 7.1.2.1 Velocity at a Point in a Channel

The flow in an open channel is generally turbulent and the velocity is observed to increase from zero at the channel bed to greater values above. A typical example is shown in Figure 7.6, with data points extracted from the observations of flowing water in a laboratory hydraulic flume. These measurements may be used in determining the total discharge passing through the channel by numerically integrating the discrete velocity values over the depth of flow, as explained further on in this chapter.

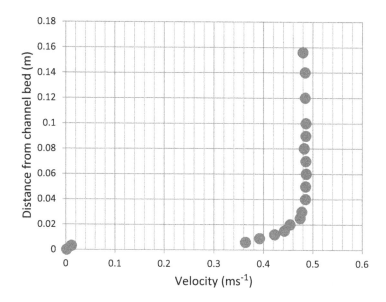

**FIGURE 7.6** Velocity measurements against depth at a station for an open channel flow experiment conducted in a laboratory. These time-averaged velocities are measured using an Acoustic Doppler Velocimeter (ADV). (Color image available in eBook).

## 7.1.2.2 Elevation Versus Reservoir Capacity

In dam engineering, estimating the volume of a reservoir is important for determining, among others, the height of the dam. The relation between the volume of a reservoir and varying heights of the dam may be used for finding an inverse relation as well. Since the reservoir volume depends upon the level of water stored, finding the volume requires first calculating the water spread area corresponding to different elevation levels. Following this, the incremental reservoir volume between any two consecutive elevation levels is computed which are then summed up for obtaining the cumulative reservoir volume as a function of elevation. Finding the areas at each elevation level may be done with the help of elevation contour maps or satellite imageries. One way for doing this is by picking up the coordinate points along the perimeter of the closed polygon enclosing each areal extent and then using a geometric formula to compute the enclosed area, a procedure that is discussed subsequently in this chapter. In the final outcome, a set of data is generated relating elevation levels and corresponding areas, as shown in Figure 7.7 The corresponding data, adapted from Mukherjee et al. (2007), is given in the table below:

| Elevation (m) | 164 | 166 | 168 | 170 | 172 | 174 | 176 | 178 | 180 | 182 | 184 | 186 | 188 | 190 |
|---|---|---|---|---|---|---|---|---|---|---|---|---|---|---|
| Area (million m$^2$) | 0 | 17 | 37 | 61 | 85 | 121 | 160 | 216 | 234 | 262 | 273 | 335 | 435 | 543 |

Converting these data into a relation between elevation and the volume of reservoir requires the use of a numerical integration technique, described further on.

## 7.1.2.3 Variation of Sediment Concentration with Depth in a Channel

Natural channels commonly carry sediment suspended along with the flow, although its concentration varies according to the sediment load that is being fed and the hydrodynamic features of the flow. Along a vertical section within the flow, the concentration of sediment is usually low nearer to the surface, increasing gradually towards the

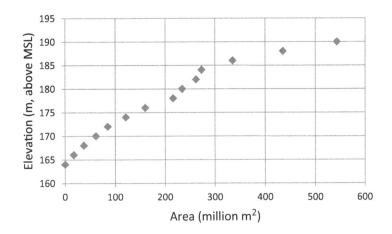

**FIGURE 7.7** Water-spread of the Hirakud reservoir, India, plotted against elevation (data from Mukherjee et al., 2007). (Color image available in eBook).

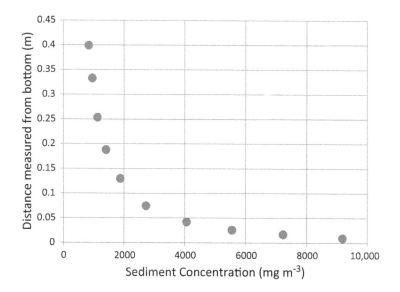

**FIGURE 7.8** Variation of sediment concentration with depth of flow in a laboratory open channel flow experiment. (Color image available in eBook).

channel bed. The measured values of sediment concentrations at different depths in a typical laboratory flow channel injected with sediment at its inlet are plotted in Figure 7.8. The data are useful in determining the total sediment load being conveyed by the flow by applying an appropriate numerical integration algorithm discussed subsequently.

### 7.1.3 AREA OF A CLOSED POLYGON

It is common in water resources engineering to calculate the area of a closed region; for example, finding the area of a catchment, the area enclosed by an isohyet, or the water-spread area of a reservoir (Section 7.1.2.2). Traditionally, this has been done using an instrument called the "planimeter" and a physical map of the area. Presently, however, with a digital version of the map it is possible to find the area using an inbuilt command of any Geographical Information System (GIS) software package. In this section, we discuss one of the mathematical methods that may be used in finding the area even if a GIS package is not available. All that would be required is a set of coordinate data, preferably in the rectangular Cartesian coordinates, for points located along the perimeter of the closed region. The set of points thus define a closed polygon, the area of which may be calculated using a simple geometric formula explained in the following section.

As an example, we choose the representative topographic map (Figure 7.9) published by Survey of India. A red straight line marks the possible location of a dam and the perimeter of the reservoir behind it is outlined with a blue continuous line that follows the elevation contour corresponding to the maximum achievable level of the reservoir behind the dam.

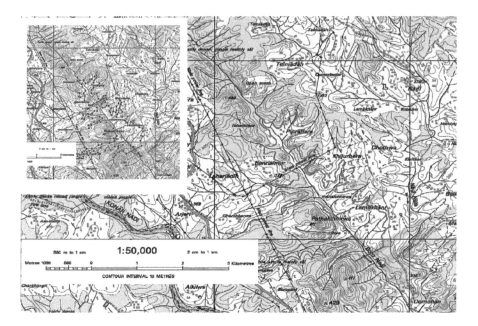

**FIGURE 7.9**    A representative region of a map showing the possible location of a dam (red line) and the perimeter of the maximum reservoir extent (blue contour-following line). Inset image at top left shows the digitization points in red. (Color image available in eBook).

For picking up the points along the contour line, we take the help of a digitizer, many online versions of which are readily available for public use. A digitizer is an application that is used to pick up the coordinate points, manually or automatically, from an image. One of these is the WebPlotDigitizer by Ankit Rohatgi (https:// automeris.io/WebPlotDigitizer/). Using the software, the $(x, y)$ coordinates of several points along the perimeter of the area under consideration are digitized (see the inset image in Figure 7.9, with the digitization points shown as red dots) and then saved in a file for finding the enclosed area. Note that the scale of the map as shown in the figure is essential during the process of digitization since it helps in correlating the physical distance on the ground to the digital pixel-units on the screen.

## 7.2   SOLUTION TECHNIQUES

Section 7.1 demonstrates a few examples encountered in hydrology, hydraulics, and water resources engineering which are described by a set of data in space or time. In this section, appropriate techniques for interpreting these data are presented. The Python codes for these techniques are described subsequently in Section 7.3.

### 7.2.1   INTERPOLATION

The method of interpolation is used when the value of an intermediate point is required to be known from a given a set of data points. This may be explained, for example, by the data of temperatures recorded at specific times of a day (Section 7.1.1.1) and the temperature at some time in between the period of observations, but not coinciding with any data point, is sought. The method of interpolation assumes that there exists a functional relation between the independent and the dependent variables. We may use, in this case, either the Newton's method or the Lagrange's method of interpolation. We discuss here the latter, as it is easier to convert the algorithm into an equivalent computer program. However, Newton's method has other advantages and the interested reader may find out about the details from standard textbooks on numerical methods, like Chapra and Canale (2021).

For "$n+1$" measured values whose independent variable is represented by $x$ and the dependent variable as $y$ and numbered from 0 through $n$, we may consider a set of data points as given in the following table.

| Independent variable, $x$ | $x_0$ | $x_1$ | $x_2$ | ... | ... | $x_i$ | ... | $x_n$ |
|---|---|---|---|---|---|---|---|---|
| Dependent variable, $y$ | $y_0$ | $y_1$ | $y_2$ | ... | ... | $y_i$ | ... | $y_n$ |

By using the interpolation technique, we shall be able to obtain a function $f_n(x)$, the expression of which is given by:

$$f_n(x) = \sum_{i=0}^{n} L_i(x) y_i \qquad (7.1)$$

In the above expression, the subscript "$n$" signifies that the interpolating function $f_n(x)$ is an $n^{\text{th}}$ order polynomial function. The function $L_i(x)$ is given as below:

$$L_i(x) = \prod_{j=0;\, j \neq i}^{n} \frac{x - x_j}{x_i - x_j} \qquad (7.2)$$

The polynomial function may be understood with the help of an example: consider a set of 4 pairs of data points $(x_0, y_0)$, $(x_1, y_1)$, $(x_2, y_2)$, and $(x_3, y_3)$ being known, implying that a 3-degree polynomial may be formed by the Lagrange Interpolation method. The expression for this third-order polynomial would have a form similar to the one shown below:

$$f_3(x) = \frac{(x - x_1)(x - x_2)(x - x_3)}{(x_0 - x_1)(x_0 - x_2)(x_0 - x_3)} y_0 + \frac{(x - x_0)(x - x_2)(x - x_3)}{(x_1 - x_0)(x_1 - x_2)(x_1 - x_3)} y_1$$
$$+ \frac{(x - x_0)(x - x_1)(x - x_3)}{(x_2 - x_0)(x_2 - x_1)(x_2 - x_3)} y_2 + \frac{(x - x_0)(x - x_1)(x - x_2)}{(x_3 - x_0)(x_3 - x_1)(x_3 - x_2)} y_3 \qquad (7.3)$$

The polynomial obtained in Equation 7.3 may be used for finding the desired value of the dependent variable corresponding to a given value of the independent variable.

The variation of the water-spread area of a reservoir with elevation, an example of which is plotted graphically in Figure 7.7, may be expressed by a polynomial function in terms of the elevation value. Similarly, the variation of sediment concentration in a flow (Figure 7.8) may be expressed as a polynomial function of the distance measured from the bottom of the channel.

## 7.2.2 REGRESSION

This method is employed when it appears that there exists a general relation, which may not be an exact function, between the independent and dependent variables for a given set of data-pair values. Also, the data points are suspected to be subject to a possible amount of error which precludes the use of the polynomial interpolation method that would thread the function through each data point, whether correct or wrong. Here we may use a regression analysis as, for example, when the water levels at a river gauging station are to be correlated to the corresponding discharges for estimating the discharge from a given water level.

In this case too, we assume a set of data pairs $(x_i, y_i)$, as used for the method of interpolation. The index $i$ denotes a general data point, which we shall assume to vary from 1 to $n$. Note that for convenience $n$ measured values are assumed unlike the $n+1$ number of points discussed previously for interpolation. We shall use the method of least-square regression, which attempts to fit an assumed function through the given points, but not necessarily coinciding with all data points, such that the sum of the squares of the residual error between the measured values of the variable $y_i$ and those calculated by the function is a minimum. The simplest relation that we may assume to exist between the variables would be a linear equation of the following form:

$$y_i = c + b \cdot x_i + e_i \tag{7.4}$$

The above equation states that for a given input value of $x_i$, the output would be the approximate value estimated by the model, $y_i$. In Equation 7.4, which is a mathematical model represented by a straight line that fits the dataset best, $c$ and $b$ are coefficients which need to be estimated from the set of given data. The residual error $e_i$ is the difference between the value predicted by the model and the true value as given by the observations.

Following Chapra and Canale (2021), the constants $c$ and $b$ in the regression equation may be evaluated by the least-square optimization method. The final values of these coefficients are as given below:

$$b = \frac{n \sum x_i \cdot y_i - \sum x_i \sum y_i}{n \sum x_i^2 - \left(\sum x_i\right)^2} \tag{7.5}$$

$$c = \bar{y} - b \cdot \bar{x} \tag{7.6}$$

In Equations (7.5) and (7.6), $\bar{y}$ and $\bar{x}$ are the average values of $y_i$ and $x_i$, respectively. Since the mathematical model relating the two variables is an approximate linear trend-line through the data points, a measure of fitness of the given data to the model needs to be defined. The commonly used metric is the correlation coefficient ($r$), defined as below:

$$r = \frac{n \sum x_i \cdot y_i - \sum x_i \sum y_i}{\sqrt{n \sum x_i^2 - \left(\sum x_i\right)^2} - \sqrt{n \sum y_i^2 - \left(\sum y_i\right)^2}} \qquad (7.7)$$

In general, a value of $r$ close to 1 indicates a good fit of the data to the assumed equation. However, the linear regression formula assumes that the variables $x$ and $y$ hold a linear dependency, which may not always be the case. For example, the data points plotted in Figure 7.3 between the water level at a gauging station and the measured discharge do not apparently show a linear relationship between them. A similar observation is applicable to all the examples of data plotted in Figures 7.5 through 7.8. Nevertheless, we may still use the linear approximation between the variables by converting the original "non-linear" relation of the data into a "linear" one by a suitable transformation. Chapra and Canale (2021) mention three different types of transformations, from which one may be chosen that matches the transformed variables as closely as possible to a linear relation.

For the purpose of demonstration, we consider the water level or gauge ($G$) of a river and its discharge ($Q$) data as shown in Figure 7.3, which is known to follow a relation such as that given below (Subramanya, 1994):

$$Q = \alpha \left(G - \beta\right)^\gamma \qquad (7.8)$$

The above equation, which is non-linear, may be linearized by taking natural logarithm ($\log_e$) on both sides of the equality, resulting in:

$$\log\left(Q\right) = \gamma \log\left(G - \beta\right) + \log\left(\alpha\right) \qquad (7.9)$$

Equation (7.9) is equivalent to writing an equation of a line in the form $y = b\,x + c$, where $y = \log (Q)$ and $x = \log (G - \beta)$. The intercept of the line on the y-axis, gives the value of $\log (\alpha)$, from where we can compute the coefficient $\alpha$ in Equation (7.8). The value of the constant "$\beta$" represents a gauge value for which the discharge is zero. This is estimated by trial and error by assuming different values and accepting the one for which the value of the correlation coefficient ($r$) is a minimum. For the data shown in Figure 7.3, by plotting the points on a double-logarithmic scale (Figure 7.10), we find that the value of $r$ is a minimum when the value of "$\beta$" is selected as 0.2 m.

Considering another example, an equation for the infiltration capacity of a soil may be obtained from the data of infiltration rates with time (Figure 7.5) by assuming a suitable relation. One of the commonly used relations is the Horton's infiltration equation, as given below (Subramanya, 1994):

$$i = f_c + \left(f_0 - f_c\right)e^{-k \cdot t} \qquad (7.10)$$

In the above equation, $i$ is the infiltration capacity at any time $t$, and $f_0$ and $f_c$ are the initial infiltration capacity at $t = 0$, and the final steady-state infiltration rate, also sometimes called the ultimate infiltration capacity. The constant $k$ in the exponent is the Horton's decay coefficient, which needs to be determined from a set of data along with $f_0$, both not directly obtainable from the observations.

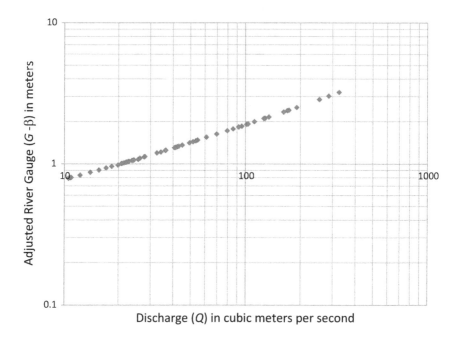

**FIGURE 7.10**  Adjusted river gauge $(G - \beta)$ versus discharge $(Q)$ plotted for the data of Figure 7.2, on logarithmic scales. Stage for zero discharge $(\beta)$ is assumed to be 0.2 m. (Color image available in eBook).

Taking logarithms on both sides of the equality of Equation (7.10), we obtain the following "linearized" equation:

$$\log(i - f_c) = \log(f_0 - f_c) - k \cdot t \qquad (7.11)$$

In Equation (7.11), which is in the form of $y = b\,x + c$, the different variables relate as: $y = \log(i - f_c)$, $x = t$, $b = k$, and the constant $c = \log(f_0 - f_c)$. By plotting the data points on a semi-logarithmic scale, with the time on normal axis, we obtain a transformed graph as shown in Figure 7.11. The slope of the graph, which is seen to be negative, gives the value of $k$ in Equation 7.11, while the line's intercept with the y-axis gives the value of $\log(f_0 - f_c)$. This helps in obtaining the unknown $f_0$ which, along with the values of $f_c$ and $k$, defines Equation 7.10 completely.

Since the value of $f_c$ is known from the original graph, the values of $f_0$ and $k$ are adjusted such that the value of the correlation coefficient $(r)$ is a minimum.

### 7.2.3  AREA-FINDING AND NUMERICAL INTEGRATION

For finding the area of an irregular closed region, as that of a reservoir behind a dam as shown in Figure 7.9, a numerical integration technique may be used that accepts a set of $(x,y)$ data points picked up along the perimeter of the region in the form given by the table on page 163. This permits us to use a numerical method for computing the area of a closed polygon defined by the selected points. The polygon is defined

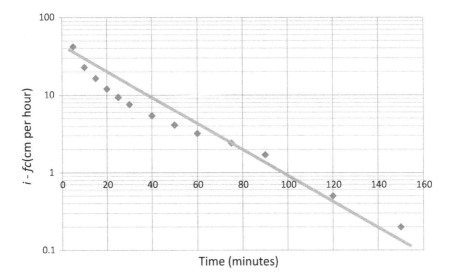

**FIGURE 7.11**    Plot of net infiltration rate $(i - f_c)$ against time $(t)$ on a semi-logarithmic scale. The inclined line is the best fit linear regression relation of the transformed data. (Color image available in eBook).

by the points when taking a round trip starting from any one of the points and coming back to it at the end of the trip. For the data as given in the table on page 163, we assume that there are "$n+1$" pairs of points, numbered from 0 through $n$.

The area of the region ($A$) enclosed by the points is then expressed mathematically as given below.

$$A = \frac{1}{2} \left\{ \begin{vmatrix} x_0 & x_1 \\ y_0 & y_1 \end{vmatrix} + \begin{vmatrix} x_1 & x_2 \\ y_1 & y_2 \end{vmatrix} + \begin{vmatrix} x_2 & x_3 \\ y_2 & y_3 \end{vmatrix} + \ldots + \begin{vmatrix} x_{n-1} & x_n \\ y_{n-1} & y_n \end{vmatrix} + \begin{vmatrix} x_n & x_0 \\ y_n & y_0 \end{vmatrix} \right\} \quad (7.12)$$

The above formula, which is sometimes also referred to as the "shoelace formula", derives its name from the cross multiplications involved for evaluating the expressions of determinants as in the following:

$$\begin{vmatrix} x_i & x_j \\ y_i & y_j \end{vmatrix} = x_i \cdot y_j - x_j \cdot y_i \quad (7.13)$$

Although not very different, other numerical integration techniques, like the trapezoidal rule or the Simpson's rule (Chapra and Canale, 2021), may be applied on the data of velocity magnitude or sediment concentration variation with depth of a channel for calculating the total discharge passing through a channel section or the total sediment load being conveyed by the channel, respectively. In these cases too, a set of data in the $(x,y)$ format, as in the following table, is made use of. It is similar to the table on page 163, but the point numbering is additionally indicated in the first row. It is assumed that the points along the $x$-axis are spaced at equal intervals and in an increasing order.

| | Point 1 | Point 2 | Point 3 | | | | Point $n$ | Point $n+1$ |
|---|---|---|---|---|---|---|---|---|
| x-ordinate | $x_0$ | $x_1$ | $x_2$ | ... | ... | ... | $X_{n-1}$ | $x_n$ |
| y-ordinate | $y_0$ | $y_1$ | $y_2$ | ... | ... | ... | $Y_{n-1}$ | $y_n$ |

The area $(A)$ enclosed by the continuous-broken line joining the ordinates $y_0$, $y_1, ...y_n$ and the $x$-axis is evaluated by the trapezoidal formula that is expressed as:

$$A = \frac{\Delta x}{2}\left\{ y_0 + 2\sum_{i=1}^{n-1} y_i + y_n \right\}$$

(7.14)

In Equation (7.14), $\Delta x$ represents the difference between the equally spaced $x$-coordinates. The Simpson's methods for finding area from the given data points are more accurate than that given by the trapezoidal formula (Chapra and Canale, 2021). The simplest of the Simpson's formula is the one-third rule and the corresponding expression for the area computed by the same set of data points given above is as follows:

$$A = \frac{\Delta x}{3}\left\{ y_0 + 4\sum_{i=1}^{n-1,\text{for odd indices}} y_i + 2\sum_{i=2}^{n-2,\text{for even indices}} y_i + y_n \right\}$$

(7.15)

One requirement for implementing the Simpson's one-third formula to the computation of area between the data points and the $x$-axis is having an odd number of points. Again, in both the methods shown, the spacing between the data points is assumed to be equal. However, quite often, the measured values of a variable are spaced unequally, as in Figures 7.6 and 7.8, showing the velocity and sediment concentration values at different depths in a stream. In such cases, the trapezoidal formula may be modified in the following way:

$$A = \Delta x_1 \frac{y_0 + y_1}{2} + \Delta x_2 \frac{y_1 + y_2}{2} + ... + \Delta x_n \frac{y_{n-1} + y_n}{2}$$

(7.16)

In Equation (7.16), the unequal spacing between two consecutive data points, say, $i-1$ and $i$ is denoted as $\Delta x_i$. The formula also assumes that there are $n+1$ data points, with ordinates sequentially ordered from $y_0$ to $y_n$.

## 7.3   PYTHON PROGRAMS

The Python codes for the different data modelling solution techniques discussed in Section 7.2 are provided here, which may be modified by the readers to suit their problems.

## 7.3.1 INTERPOLATION

The Lagrange's interpolation method for a set of $n + 1$ data pairs in the form $(x_0, y_0)$, $(x_1, y_1)$, $(x_2, y_2)$ ... $(x_n, y_n)$ may be coded into Python as given below. A sample data set is included for the squares of the first five natural numbers. When the program is run, the user is prompted to enter the value of $x$, which may be any real number between the two extreme x-values, and the corresponding value of the square of the number is returned on the screen.

```
# Lagrange interpolation
import numpy as np
x=np.array([0,1,2,3,4])
y=np.array([0,1,4,9,16])

xx = float(input("Enter x:"))
n = len(x)
sum = 0
for i in range(n):
    product = y[i]
    for j in range(n):
        if i != j:
            product = product*(xx - x[j])/(x[i]-x[j])
    sum = sum + product
print('For x = %.2f, y = %.5f' % (xx,sum))
```

The same program may be used to find the intermediate value between two data points in Figures 7.1 to 7.8, by replacing the x- and y-arrays with the relevant data sets. It must be remembered, however, that in any of these data sets the data pairs are not related by a functional relation, as was for the example of natural numbers and their squares. Hence, the polynomial interpolation function may not provide a good result if all the data pair values from a data set are entered. For best results, one has to choose only a few data pairs in the neighbourhood of the point under consideration. For example, considering the water-spread areas of the reservoir plotted against elevation in Figure 7.7, assume that it is required to find the area corresponding to a specific elevation that lies between 160 and 170 m. Since this corresponds to the first four datapairs of the fourteen given data-pair values, we replace the program's input data with the following x- and y-arrays as shown below:

```
x=np.array([164,166,168,170])
y=np.array([0,17,37,61])
```

The x-values represent the elevations (in meters) and the corresponding y-values are the water-spread areas (in million m²). On running the program, the user is prompted for an elevation value, and the program computes and returns the interpolated value of area. However, the result is likely to degrade gradually as more and more points are included in the interpolation program since the polynomial, being forced to weave through all the data points, may start showing oscillations unless the points line up perfectly along the polynomial's trajectory. Hence, in such cases, it would be a better choice to find out a trend line using the method of regression, as given in the following section.

## 7.3.2 Regression

The Python code for implementing the regression formula (Equation 7.4, with the coefficients defined in Equations 7.5 and 7.6) is given below. The sample data of infiltration rates shown in Figure 7.5 and in the linearized form plotted on a semi-logarithmic scale in Figure 7.11, are chosen for demonstrating the code. Since a linear regression line is required that passes through the transformed data points (Figure 7.11) best, the sample data included in the Python program comprise the series of observation times as the $x$-array, and the corresponding natural logarithms of $f_0 - f_c$ as the $y$-array. As mentioned before, $f_0$ and $f_c$ are the infiltration capacities at the beginning and at a time when it has reached a steady state.

```
# Regression analysis
import numpy as np
x=np.array([5,10,15,20,25,30,40,50,60,75,90,120,150],float)
  y=np.
array([3.74,3.13,2.80,2.48,2.24,2.03,1.69,1.41,1.16,0.87,0.53,
  -0.69,-1.60],float)
x_mean = np.mean(x)
y_mean = np.mean(y)
m = len(x)
numer = 0
denom = 0
for i in range(m):
    numer += (x[i]  -   x_mean) * (y[i]  -   y_mean)
    denom += (x[i]  -   x_mean) ** 2
b = numer / denom
c = y_mean - (b * x_mean)
print('b = %.3f, c = %.3f' % (b,c))

ss_t = 0
ss_r = 0
for i in range(m):
    y_pred = c + b * x[i]
    ss_t += (y[i] - y_mean) ** 2
    ss_r += (y[i] - y_pred) ** 2
r2 = 1 - (ss_r/ss_t)
r = np.sqrt(r2)
print('r = %.3f' % (r))
```

When the code is run with the given data, we obtain $b$ as −0.033 and $c$ as 3.280. The correlation coefficient ($r$) is found to be 0.986, which is quite good. On comparing with Equation 7.11, we obtain $\log (f_0 - f_c) = 3.280$. Since $f_c$ was found to be about 6.0, we may find $f_0$ by writing this equation as $(f_0 - 6.0) = e^{3.280}$, which gives $f_0 = 26.58$. The value of $k$, which is the same as $b$, equals −0.033. Hence, Equation 7.11 may be written by substituting the now known values of the different parameters as:

$$\log(i - 6.0) = \log(26.58 - 6.0) - 0.033t \qquad (7.17)$$

Equation 7.17 may also be written in the form of Equation (7.10) as under:

$$i = 6.0 + 26.58\,e^{-0.033t} \qquad (7.18)$$

By using Equation 7.18, and by varying the time ($t$) from 1 to 120 s, we obtain the modelled results which are plotted in Figure 7.12, along with the measured data. Barring a few intermediate points, the rest of the model predictions appear to match quite well with those observed. Of course, this is a demonstration of just one of the infiltration models – the Horton's equation – and the same methodology may be applied to choose any other model that fits the data best. For details about other infiltration models, relevant texts, such as Subramanya (1994), may be referred to.

Another example of application of the regression method is to find a (negative) correlation between two parameters, such as that of temperature and humidity plotted in Figure 7.1. As mentioned before, a visual observation indicates an apparent inverse relationship between the two variables. This becomes clear when the two variables are plotted one against the other as shown in Figure 7.13. A linear mathematical relation of the form $y = bx + c$ may be proposed in this case, where the $x$-variable is equated to temperature and the $y$-variable, the corresponding humidity.

On running the program with the data pairs of temperature and humidity for the data plotted in Figure 7.1, the value of the coefficient $b$ is obtained as −3.168, while that of the constant $c$ as 180.3. The correlation coefficient ($r$) is obtained as 0.963, which is acceptable, although an amount of scatter is noticeable for the points around the line of best-fit shown in the figure.

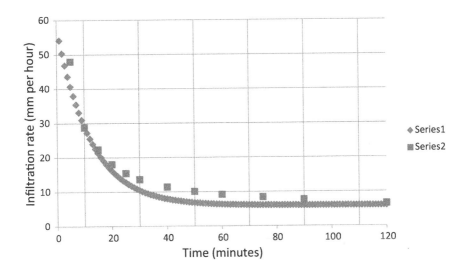

**FIGURE 7.12**  Comparison of infiltration rates by model predictions using Equation 7.18 and observed values as reported by Adindu et al. (2013). (Color image available in eBook).

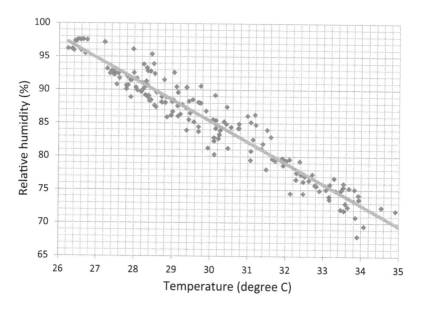

**FIGURE 7.13** Scatterplot of the temperature versus humidity data, which were shown as time-series plots in Figure 7.1. The inclined line is the best fit linear regression relation passing through the data points. (Color image available in eBook).

## 7.3.3 COMPUTATION OF AREA

The reservoir water-spread area behind a proposed dam shown in Figure 7.9 is digitized by a sufficient number of points (Figure 7.9, inset, showing the red dots as the digitization points) and saved as a data file in the $(x,y)$ format. The file may be in the .csv (comma-separated values) format or any other, like the space-, or tab-separated formats, and saved with a .txt file extension. While digitizing, the geographical scale shown in the figure is made use of for calibration, that is, for providing a correlation between the digitized pixel-points on the screen to the actual locations on the ground with respect to an assumed origin. For the purpose of demonstration, we presume the file "input.csv" is available after carrying out the digitization process and we use it in the sample Python code for computation of the area enclosed within the points, as shown below.

```
# Computation of area of a closed polygon
import numpy as np
coords = np.loadtxt("input.csv", delimiter=",")

def PolygonArea(c):
    n = len(c) # of digitization points
    area = 0.0
    for i in range(n):
        j = (i + 1) % n
        area += c[i][0] * c[j][1]
        area -= c[j][0] * c[i][1]
    area = abs(area) / 2.0
```

```
    return area

area = PolygonArea(coords)
print("area enclosed by points = ",area)
```

On running the program, an enclosed area defined by the points along the reservoir periphery is found out as 3,059,900 m², that is, just about 3 km². The same technique may be extended to compute the water spread area of a reservoir at various elevations (and represented graphically, for example, as the data plotted in Figure 7.7), from the corresponding surveyed maps or satellite imageries.

### 7.3.4   NUMERICAL INTEGRATION

The numerical integration method using the trapezoidal rule with unequal data spacing is demonstrated with a Python code for computing the discharge per unit width in a channel for the observed velocity profile shown in Figure 7.6. The code uses Equation 7.16, but the $x$- and $y$-coordinates of the equation are interchanged for this problem since the velocities are reported in terms of the vertical distance, $y$, measured from the bottom of the channel. Thus, in the Python code, $x$ represents the velocities (in ms⁻¹) measured at a certain level $y$ (in m). Also, note that integrating the velocities over the depth of flow multiplied by 1-m width of the channel gives the "unit" discharge, or the discharge per unit width.

```
# Numerical integration
import numpy as np
x=np.array([0.0017,0.0114,0.3638,0.3922,0.4226,0.4425,0.4539,
   0.4743,0.4776,0.4852,0.4855,0.4866,0.486,0.4827,0.486,0.4862,
   0.4846,0.4847,0.4801],float)
y=np.array([0,0.003,0.006,0.009,0.012,0.015,0.02,0.025,0.03,
   0.04,0.05,0.06,0.07,0.08,0.09,0.1,0.12,0.14,0.156],float)

m = len(x)
sum = 0
for i in range(2,m):
    dy = (y[i] - y[i-2])
    prod = dy * x[i-1]
    sum += prod

sum += (y[1] - y[0]) * x[0]
sum += (y[m-1] - y[m-2]) * x[m-1]
print('sum = %.3f' % (sum))
```

On running the above code, the unit discharge in the channel is computed as 0.144 m³s⁻¹ per m.

A further example of the method of numerical integration may be demonstrated in the context of the elevation versus water-spread area curve for a reservoir shown in Figure 7.7. Each of the areas corresponding to an elevation is computed by the technique discussed in Sections 7.2.3, which is converted into a Python working code in Section 7.3.3. However, in dam engineering, a more useful graph than the elevation–area curve is the elevation–volume (or elevation-capacity) curve for the reservoir. This graph plots, in place of water-spread area, the volume of the reservoir corresponding to an elevation. The volume of the reservoir enclosed between two consecutive elevation contours may be computed by applying a numerical integration technique, like the trapezoidal formula, Equation 7.14. The cumulative volume of the reservoir corresponding to an elevation may be found by adding all the individual volumes between the lowest point of the reservoir and the given elevation. A program in Python for the purpose is given below.

```
# Trapezoidal rule integration for reservoir capacity
import numpy as np

x=np.array([164,166,168,170,172,174,176,178,180,182,184,186,
  188,190],float)
y=np.array([0,17,37,61,85,121,160,216,234,262,273,335,435,
  543],float)
dx = 2.0
m=len(x)

volume = np.zeros(m)
for i in range(m-1):
    dvolume = 0.5*(y[i]+y[i+1])*dx
    volume[i+1] = volume[i]+dvolume

print(x,volume)
```

The initial lines of the program provide the necessary data, such as the elevations (in m) and the corresponding water-spread areas, $y$ (in million m$^2$). On running the program, the series of elevations and corresponding reservoir capacities are printed. These values, when plotted in a graph, look similar to that in Figure 7.14.

We may also use the Simpson's numerical integration formula discussed in Section 7.2.3. The corresponding formula is given in Equation 7.15 but the application requires having an odd number of data values. However, in the present case of finding reservoir volume with elevation, there are 14 such points given as shown in the Python program above. Hence, we use 13 of these points, leaving out the first which may be computed separately and added. The program in Python for finding the reservoir capacity between elevations 166 m and 190 m is given below.

```
# Simpson's formula for finding reservoir capacity
import numpy as np

x = np.array([166,168,170,172,174,176,178,180,182,184,186,188,
  190],float)
y = np.array([17,37,61,85,121,160,216,234,262,273,335,435,
  543],float)
dx = 2.0
m = len(x)
```

```
volume = y[1]
for i in range(1,m-2,2):
    volume += 4*y[i]+2*y[i+1]
    print(i)
volume += 4*y[m-2]+y[m-1]
volume = volume*dx/3

print ("%3.2f"%volume)
```

On running the program, the volume of the reservoir between 166m and 190m is computed, leaving out the volume between the first elevation and the reservoir bottom, and is found out as 4977.3 million m³. The part that is left out at the bottom of the reservoir is in the form of an inverted pyramid with its apex, corresponding to an elevation of 164 m, may be assigned a zero area while its base, corresponding to the elevation of 166 m, is 17 m². The following standard formula for finding the volume of a pyramid ($V_p$) with a given base area ($A_b$) and height ($h$), may be used for the calculations:

$$V_p = \frac{1}{3} A_b h \tag{7.19}$$

By substituting the relevant data, $V_p$ is obtained as 11.3 million m³ which, when added to the volume of the upper portion of the reservoir (4977.3 million m³), gives a total of 4988.6 million m³. This value is slightly lower than that computed for the total reservoir capacity by the trapezoidal method (Figure 7.14), which is 5015 million m³. The difference between the two, of course, is less than 1%. It may be remembered that the Simpson's method is more accurate than the trapezoidal method as its error is proportional to $1/n^4$ in comparison to $1/n^2$ for the latter (Chapra and Canale, 2021), where $n$ is the number of intervals of data points.

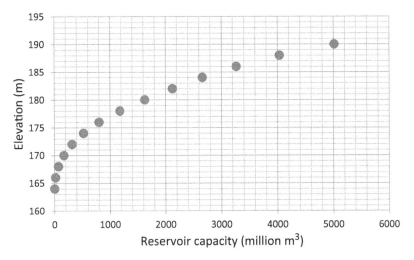

**FIGURE 7.14**  Elevation versus reservoir capacity plot computed from the data of elevation versus water-spread area shown in Figure 7.7. (Color image available in eBook).

## REFERENCES

Adindu, R. U., Igboekwe, M. U., Ughegbu, A. C., Eke, K. T. and Chigbu, T. O. (2013). "Characterization of Infiltration Capacities of the Soils of Michael Okpara University of Agriculture, Umudike – Nigeria". *Geosciences*, 3(4), 99–107.

Chapra, S. and Canale, R. (2021). *Numerical Methods for Engineers*. McGraw-Hill Education, 8th edition. https://www.mheducation.com/highered/product/numerical-methods-engineers-chapra-canale/M9781260232073.html

Mukherjee, S., Veer, V., Tyagi, S. K. and Sharma, V. (2007). "Sedimentation Study of Hirakud Reservoir through Remote Sensing Techniques". *Journal of Spatial Hydrology*, 7(1), 122–130.

Subramanya, K. (1994). *Engineering Hydrology*. Tata McGraw-Hill Education, New Delhi, India.

Martinson, D. G. (2018). *Quantitative Methods of Data Analysis for the Physical Sciences and Engineering*. Cambridge University Press, Cambridge, UK.

# Index

## A

accuracy and stability, 49
advection, 136, 143
aquifer, 112–119
area computation, polygon, 161, 166, 172

## B

back substitution, 15
backwater curve, 58
bio-chemical oxygen demand (BOD), 43, 61
boundary condition, 46–47, 72, 84, 87, 94, 96, 104, 114, 119, 128, 137, 141
boundary value problem, two-point, 51

## C

central differences, 137
channel contraction, 7, 23
channel hump, 7, 23
confined aquifer, 113
confined flow, in aquifers, 113
contaminant transport, 45, 70, 133, 148
Courant number, 139
critical depth, 40
critical flow, 24

## D

Darcy law, groundwater flow, 111
Darcy–Weisbach equation, 11, 44
data
    flow velocity, 159
    humidity, 154
    infiltration, 157
    reservoir capacity, 160
    river discharge, 155
    river stage, 155
    temperature, 154
depth-averaged flow, 78, 86, 104
design flood hydrograph, 38
diffusion, contaminant, 135, 136, 141, 143
direct runoff hydrograph, 30
Dirichlet boundary condition, 46, 66, 71, 137
dispersion, contaminant, 46, 71, 134, 136–137, 146, 148
dissolved oxygen (DO), 41, 61
dissolved oxygen deficit, 41
Dupuit assumption, 44

## E

emptying of tank, 37
energy line, 10
Euler's method, 47

## F

finite differences, 4, 51, 66, 71, 79, 81, 83, 84, 87, 114–119, 137
first-order ordinary differential equation, 47
flood hydrograph, 81
flood routing, 38, 54, 81, 91
forward differences, 137, 139
free surface flow, 76
friction factor
    channels, 5, 79
    pipes, 44, 65

## G

Gauss elimination method, 4, 10, 15
gradually varied flow, 39, 58
groundwater
    flow, 44
    table, 44, 66

## H

head loss, 11, 42, 44
Heun's method, 48, 61, 64
hydrograph, 54, 82, 94

## I

ideal fluid flow, 78, 84
inflow hydrograph, 38, 93
initial conditions, 39, 44, 47, 49, 51, 87, 96, 120
interpolation, Lagrange, 162, 169
iterative solution, 14, 19, 24, 31, 34, 48, 49, 56, 69, 99

## K

kinematic wave equation, 79, 81, 91, 94

## M

Manning's equation, 79
Manning's friction coefficient, 5, 18, 58, 60, 79, 89, 96, 104, 108